AF890204

Motorcycle Modification

Keeping your Special Safe

by

John Hardaker

ISBN: 978-1-916732-05-6

Second Edition - 2024

Copyright 2024

All rights reserved. No part of this publication may be reproduced, stored in a retrieval system or transmitted in any form or by any means, electronic, mechanical, photocopy, recording or otherwise, without prior written consent of the copyright owner. Nor can it be circulated in any form of binding or cover other than that in which it is published and without similar condition including this condition being imposed on a subsequent purchaser.
The right of John Hardaker to be identified as the author of this work has been asserted in accordance with the Copyright Designs and Patents Act 1988.
A copy of this book is deposited with the British Library

Published By: -

i2i
PUBLISHING

i2i Publishing. Manchester.
www.i2ipublishing.co.uk

Contents

Motorcycle Modification

Keeping your Special Safe

Introduction

Following a lifetime interest in motorcycles and all things mechanical, and 50+ years in the aircraft and aero engine industry, it seems like a good time, having passed the age of 80, to record some of the lessons learned in this time. Things which are perhaps not normally part of an engineering education, but which can happen in the real world and could have unfortunate consequences.

I hope the book will be of interest to motorcycle special builders and all those interested in making changes to manufacturer's standard specifications and perhaps to students of engineering and those just generally interested in motorcycles. To any engineers who may be among the readers, some of the points made may seem like an exercise in stating the obvious but these things are only obvious if you are familiar with them, and I have tried to make the book understandable to all motorcyclists and potential special builders. Few things are more satisfying than riding out on a machine of your own design and construction, something different from the general run of production machines.

I have based the book on the techniques of failure analysis which are part of the certification requirements in the aircraft industry, (although for some unknown reason they have changed its name and now call it Safety Analysis). Whatever we call it, its use is one of the main reasons why accidents as a direct result of mechanical failure are very rare in aviation and when they do occur, there is often some element of human error in the story.

It is hard to argue that there is not some element of risk in riding a motorcycle, travelling at speed through a mass of stationary and "randomly" moving solid objects, on surfaces which can change unexpectedly, but overcoming these difficulties is part of the attraction. If it was totally safe and boring, it would not be worth doing. However, all the skill, care, attention and experience which may be applied to riding is of no account if the bike fails to consistently go where it is pointed or does not consistently accelerate or stop when required.

This satisfactory and predictable running can only be achieved by attention to detail in the design, manufacture and maintenance to ensure that nothing which could cause a problem when riding is overlooked and in particular ensuring that all likely failures, which cannot be eliminated by design, do not result in loss of control, or if that is unavoidable, that they have an acceptably low probability of occurrence.

Although a very powerful tool for ensuring safety, a formal detailed failure analysis of the type used in the aircraft industry, is usually very long winded and large parts of it are quite tedious to prepare, but just occasionally you find a small point which has been overlooked and which, if it resulted in a failure, could cause a real problem. Think of a small insignificant part which, if it failed or came loose, could lock the steering or lock a wheel on a motorcycle. Once found, this type of problem can usually be designed out fairly easily but the best time to identify them is not at 70 mph in the middle of a corner.

It is not intended, in this book, to attempt any sort of full analysis, which in any case would be different for each individual design. I hope I can show the principles of the technique and how it can be used to review any proposed change in advance to ensure that there are no unfortunate consequences.

To provide a bit of light relief for the reader I have included various reminiscences from both 2 and 4 wheels and from the aircraft industry which I hope are of interest and illustrate the sort of thing which can happen. As a general rule, anything which can go wrong inevitably will, and it usually happens at the most inconvenient moment.

Because my motorcycling experience began in the 1950s (and still continues) some of the examples are from bikes made before many current motorcyclists were born. This does not make them any less relevant today. The basic principles of good design remain the same and, if anything, there are more ways of getting it wrong today that there were in the past with the increased complication of modern bikes and the addition of electronics.

Finally, I hope the book will encourage more motorcyclists not to be frightened of doing some modifications. If there is something you don't like about your machine, don't live with it, try and think how it could be improved. If you come up with a good idea, make or buy the necessary parts and try it out. You have nothing to lose, providing that you look after the safety aspects (and avoid invalidating any warranty if it is a new bike). Hopefully, the suggestions in this book will help with ensuring that the modification is safe.

For those with no facilities to make parts, there are plenty of small companies who will make almost anything you need for reasonable money and also an endless variety of parts available from breakers and auto jumbles which can often be adapted to fit. It is very satisfying if the modification works and improves the bike but, if not, you can always go back to standard. Success might encourage you to do more modifications or perhaps even build a unique bike.

Part 1 of the book is a general introduction to the principles and purpose of failure analysis. This part may seem a bit remote from motorcycles, but I hope you will see the relevance to all things mechanical where failures could result in a dangerous situation.

Part 2 is included to show a selection of my efforts over the years which may encourage anyone thinking of making some changes to their machine.

Part 3 looks at particular parts of a motorcycle and suggests which aspects should be looked at when doing modifications or designing and making new parts for a motorcycle.

I have included Part 4 as an attempt to look into the future, probably as wrong as most people's attempts, but one thing is certain. The basic failure analysis method will continue to be the best way of ensuring that all products are safe to use.

When the opportunity to update the book with a second edition arose it seemed a good time to add a fifth section to pass on my recent experience converting two wheeled bikes into three wheelers. This started mainly just to keep myself mobile, being well past my best on two wheels. As I got more involved in the detail, I soon realised that there is a great difference between just making the bike stand on three wheels and making something which handles well on the road and is a pleasure to ride.

Although somewhat outside the original purpose of the book, Section 5 is an attempt to pass on things I have learned about this type of conversion in the hope that it may help anyone thinking of something similar.

PART 1

What is Failure Analysis?

Chapter 1 – Principles

The first thing to say about failure analysis is that it is not the same as failure investigation.

Failure investigation is work done to find the cause of a failure or accident after it has happened, generally by reconstruction of the circumstances before the accident and inspection of the failed parts to find the true cause.

Failure analysis is work done during design and construction to ensure a safe machine where all failures which can reasonably be expected to occur during its service life are anticipated and investigated with the aim of minimising their effects on the safety of the rider and everyone else.

You could say that it is the application of common sense in a logical manner to the possibility that any part of the machine could fail. The analysis method described in this book is the so-called "bottom up" method where every part of a machine is investigated separately to see how it could fail and what the final effects of that failure would be.

As a simple example to illustrate the principle, you could say, "If this particular bolt failed or came loose, does anything fall off, and if something does fall off or move out of position, could it cause any danger to the rider or anybody else?" If it may do so, we need to do something to prevent it ever happening.

As there are a lot of parts in a motorcycle, a full analysis is a big job but in the case of a modification or special build, only the modified parts would need to be covered.

The other main method of analysis is the so-called "top down" method where the ultimate effects which could be dangerous are selected first and the analysis attempts to find all possible causes. It is easy to list the effects of failure that we really don't want to happen when riding our creation, the main ones being:

- Loss of control of braking or steering, or of the ability to control or reduce engine power transmitted to the rear wheel.
- A permanently locked wheel at either end.
- Unexpected loss of tyre adhesion, particularly when cornering.
- Uncontrollable instability.
- Total loss of lights in total darkness.
- Anything which reduces the ability of the rider to control the machine.

The problem with the "top down" method is that, with something like a motorcycle, you tend to miss some of the more obscure causes which can subsequently catch you out. This method is better for a system with a fixed number of components, each with a known failure rate, such as an electrical circuit.

In general, the designer of a machine is not the best person to do failure analysis as they design for success and often dismiss possible failures with, "It could never happen to my brilliant design because of X, Y and Z," rather than, "What if it did?" Brilliant design or not, if anything bad can happen, it probably will at some time, so the best time to identify it is before it happens.

This reminds me of a close call in my early days of special building, before I had heard of failure analysis. In the 1950s my first big teenage motorcycle project was to modify a pre-war Excelsior Manxman to the, then relatively new, telescopic forks and swinging arm rear suspension. I was very confident that the "brilliant design" would work perfectly. (see Photos 1 and 8).

Cutting up a rare vintage machine would be frowned upon now but in the '50s it was just another old bike. As far as I remember, the front end was easy using a Matchless fork assembly and steering head races ground out by a helpful man at work. In those days, the occasional "foreigner" in aid of better cars and motorbikes was tolerated by the management as "encouraging the lads."

The rear end was more difficult with a swinging arm welded up from thick wall steam pipe, twin shocks and a bolted on rear sub-frame. The whole thing was over engineered and very strong, so I was confident.

Although I was the "designer" and had never considered the possibility of failures at that time, it must have been instinctive as most of the obvious possible failures were covered either by excessive strength or dual load paths which allowed any single failure to be tolerated without the structure collapsing.

The bike went very well at first and was able to keep the then current sports bikes in sight, although it was a bit difficult to keep the oil inside the engine. The old Manxman was not slow and certainly looked the part with its OHC engine in the days when most road bikes were push rod OHV and new side valve bikes were still being made. However, after a few weeks the flaw in the design emerged to bite us.

The torque link, which formed the rear brake anchor, was mounted below the wheel centre line and was attached to a large tight "P" clip around the swinging arm which was fitted against a stop provided by the end fitting of the swinging arm, thus preventing the anchor point moving rearwards. The first part of Sketch 1 shows the arrangement.

Again, the whole thing was excessively strong. All logic said that forces on the brake anchor torque link were either zero or in a rearward direction during braking and thus the stop would prevent the brake anchor point moving with no reliance on friction of the "P" clip around the swinging arm.

However, illustrating the point that things do not always work out as the perceived logic of the designer thinks they should, for no known reason, the "P" clip worked its way forwards along the swinging arm against the direction of the major forces, and in spite of the clamping friction of the tightened clip, allowing the brake plate to rotate forwards pulling on the drum brake operating rod, which locked the wheel. The second part of Sketch 1 illustrates the sequence of events.

Whether by luck or skill, a friend of mine rode the bike to a stop with a locked back wheel from a considerable speed and stayed upright. (He claimed 70 mph but was not known for letting the facts get in the way of a good story). We made little of it at the time, just congratulated him on his skill, welded the "P" clip to the swinging arm in the correct position and carried on, but it was a lesson learned and never forgotten, even though it was over 60 years ago. A good example of the failure of a small insignificant part causing a real problem on the road.

The rider was not discouraged by the incident and subsequently bought the bike from me and used it regularly for a number of years.

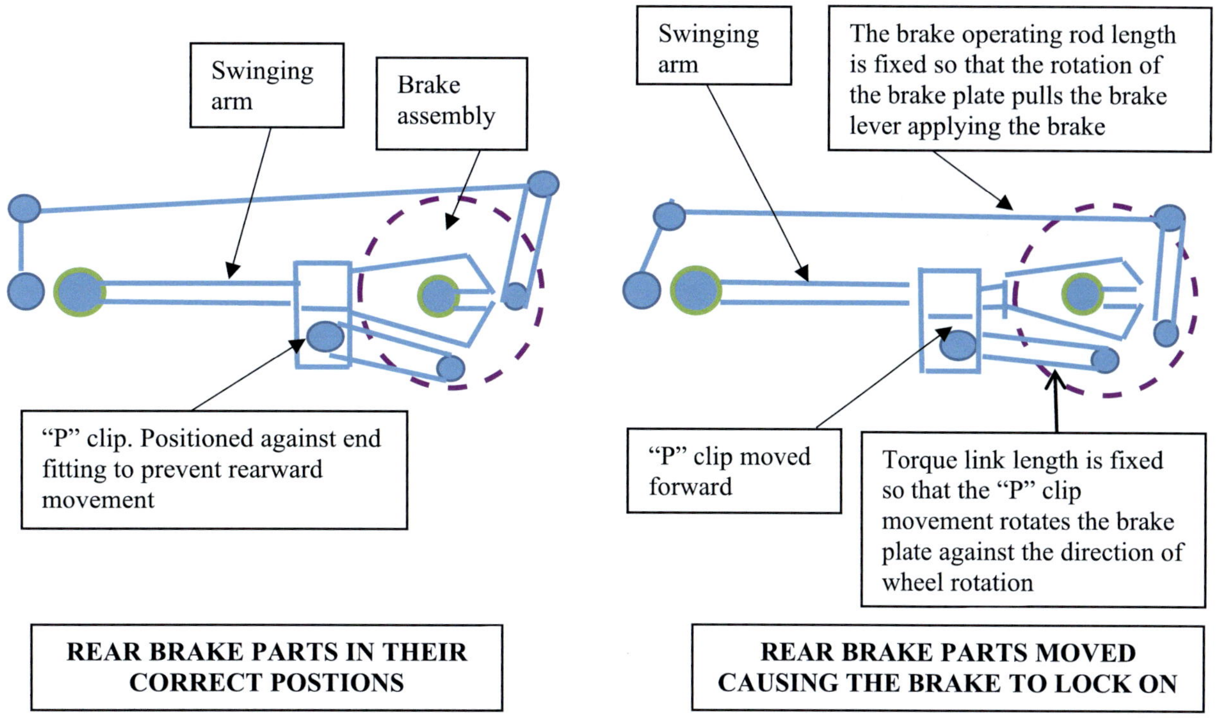

Sketch 1 – The Excelsior rear brake problem.

Thinking about it now, it seems to me that the sequence of events probably started with the wheel spindle nut and the "P" clip clamping bolt both becoming slightly loose (there would probably have been no thread locking). This would allow the brake plate to rotate backwards on the wheel spindle due to gravity, suspension movement and vibration, (its centre of gravity would be behind the wheel centre). This would push the "P" clip forward and also pull on the brake operating rod. When movement was enough to apply the brake, the "P" clip would be locked, by "cross binding" friction in its forward position, due to the offset load in the torque link, making it impossible to release the brake.

Things don't always happen as you predict that they should, and all possible failures should be considered, not just those which seem likely, in line with thinking at the time of design.

Photo 1 – The almost completed Manxman in the 1950s.
Plenty of confidence in the design but the development of the necessary caution and assessment
of risk was still "work in progress." Photo 8 in Part 2 shows more details of the Excelsior.

Chapter 2 – Types of Failure

As part of the investigation to establish where the effects of possible failures could result in danger to personnel, aircraft failure analysis usually divides the possible failure cases into categories as follows.

Various aviation authorities have different terminology, but the following is based on the requirements of the European Aviation Safety Agency (EASA), which is the body responsible for certification of all aircraft and aircraft engines designed and made in the UK and Europe. The US equivalent, the Federal Aviation Administration (FAA) has similar "Requirements" although there are detailed differences in terminology, and together these two sets of requirements cover the large majority of the world's civil aircraft.

Failures which do not affect continued flight, at worst causing something like a single engine shut down on a multi-engined aircraft, are categorised as **Minor**.

The equivalent on a motorcycle would be a failure which at worst caused the machine to roll to a stop at the roadside. Unfortunate and possibly expensive, but not much chance of danger to anyone involved. It is worth noting that these categories are nothing to do with the potential cost of repair. A severe engine failure which could write off most of the engine would still be **Minor** providing that it did not seize up, catch fire or eject high energy parts, and thus did not endanger the rider.

Failures which would result in a severe risk of death or injury to any person are categorised as **Hazardous**. For example, an uncontained engine failure where debris may be ejected through the engine casings and impact the aircraft would be classified as **Hazardous**.

The equivalent on a motorcycle would be a failure which made the machine uncontrollable with a serious likelihood of a crash. The rear wheel locking referred to in chapter 1 would definitely be categorised as **Hazardous** even though we were lucky in that case.

Between **Hazardous** and **Minor** is a category known as **Major** where the effects are not necessarily **Hazardous** but demand correct pilot actions and/or a level of flying skill to overcome the problem. A fire under an aircraft engine cowling, which would be contained within fireproof boundaries, and could be safely controlled and extinguished by pilot actions, would be an example of a **Major** failure.

The equivalent on a motorcycle would be something like an engine seizure which would lock the back wheel momentarily but could be overcome by quick action with the clutch.

The final category is **Catastrophic**, which is in effect a failure which makes the aircraft incapable of further flight, and thus will inevitably result in a crash. You could think of a wing falling off, or all engines stuck in reverse thrust, as extreme examples of **Catastrophic** failure.

The equivalent on a motorcycle would be something like the forks breaking off under the steering head.

We obviously need to ensure that **Hazardous** or **Catastrophic** failures will never happen with our design, but in order to do that, the first step must be to identify all of them and be aware of the consequences, thus taking no risks in these areas.

Again, it is just the logical application of common sense. The desirability of straightening a badly bent fork leg would obviously need to be considered in a different light from straightening a bent number plate support. A convention used in failure analysis is that the defined failure types and part categories such as **Hazardous** and **Critical** always have capital letters, so I have stuck with this in the book. Also, I have referred to any part which is unique to this machine or modification as a **one-off** part or machine, a convention used to indicate that only one set of parts to this design exists.

Chapter 3 – Design for Safety

There are two separate aspects of design which we should consider when designing modifications to our motorcycle. Obviously, the design must fulfil its function, be strong enough for its purpose and also, where visual appeal is important, must look good.

Having produced a design which fulfils the above requirements you could just make the parts, fit them, and move on the next modification. After all, your design is so brilliant there is no possibility of failure.

No doubt this conclusion will be correct in the large majority of cases but with a large number of parts in the whole motorcycle it is hard to be sure that every possible failure case has been considered, including interaction between different parts (i.e. damage to, or failure of one part resulting in a **Hazardous** failure of another adjacent or connected part), whether considered likely or not. This is where failure analysis comes in as a second stage to the design process to provide a check on the safety of the design in all conceivable circumstances.

The importance of failure analysis in aircraft design cannot be overstated and its rigorous application is one of the main reasons why air travel is safe enough to allow unrestricted travel by the general public without any real worries about safe arrival. Part of aircraft certification requires a full failure analysis of all parts with justification of safe operation in all cases. This is a very big job employing numerous engineers for something like a year. The whole thing is produced in tabulated form and the results submitted to the certification authority for approval. A simplified example covering a motorcycle fork is shown in Appendix 1 for interest.

The aircraft methods described below are a very formal process, carefully documented and therefore are not practical for a one-off motorcycle, but I have included the following summary of the aircraft method in order to highlight the many aspects of the job and the way by which we could achieve a similar result for a motorcycle. The basic methods by which safe operation can be justified, are no different. It just needs to be done mentally when deciding on details of the design.

1) ELIMINATION OF HAZARDS

By far the best way to deal with a potentially **Hazardous** or **Catastrophic** failure case is to change the design to eliminate it. This is not practical in some cases but with a bit of thought a surprising number of potential **Hazards** can be designed out.

In a typical modern aircraft engine consisting of several thousand parts there would normally be only about 20 to 30 parts classified as **Critical** (see section 2) because failure of that part would be a **Hazard** to the aircraft. The design would be arranged such that the worst effect of failure of any other part would be to shut down the engine, which is not **Hazardous** in a multi-engined aircraft. A single engine aircraft is fundamentally less safe in the case of engine failure but there is still a difference between an unpowered aircraft which is still under control as a glider and total loss of any control of flight.

For a motorcycle example, think about a rear suspension collapse as a result of failure of the top or bottom mounting of a single combined spring and shock absorber unit. If the design relies totally on an internal bump stop in the shock absorber to limit wheel movement, this would become ineffective with the failure of the shock mountings, and the rear of the bike would fall onto the tyre, locking the wheel, a definite **Hazard** (in spite of our experience with the Manxman where luck played a big part).

However, it should be possible, without much change to the design or significant extra weight, to arrange that the swing arm contacts some part of the frame, or other solid structure, as an "independent bump stop of last resort" before the tyre touches anything which may cause the wheel to lock.

This very small change, which will have no effect and be unnoticed for the vast majority of the bike's life may just save the day if one day the unexpected happens.

This is a good example of the purpose of failure analysis which is to make the machine as failure proof as possible so that whatever happens, likely or unlikely, the possibility of an accident as a direct result of a failure of the machine is minimised.

2) CRITICAL PARTS

In some cases, it is not possible to design out a **Hazard** without excessive weight or complication, so we have to live with it. This is true in aircraft as well as cars and motorcycles and these parts are known as **Critical Parts**. The method for dealing with the safety of a **Critical Part** is based on probabilities. We must accept that no type of failure is totally impossible but if the probability of failure can be made low enough there is very little chance of it happening within the life of the machine.

As an example of a single component where failure would be **Hazardous**, a turbine disc in an aircraft gas turbine engine is, in effect, a large metal disc with the turbine blades attached around the rim and rotates at very high speed. The disc is subject to high internal stresses resulting from centrifugal forces and temperature variations between the bore and the outer rim and, if it should burst, the released pieces of the very hot disc would have sufficient kinetic energy to go straight through the engine casing and cowling and impact the aircraft with enough remaining energy to go straight through; a definite **Hazard**. It would be possible to eliminate this **Hazard** by means of a very thick "armour plate" containment casing around the turbine but the weight penalty of this course of action would be totally unacceptable on an aircraft.

The aircraft certification requirements refer to this type of component as a **Critical Part** and demand that the maximum acceptable probability for a failure of this type in an aircraft engine is not more than 1 case in 100 million flight hours for the whole fleet of aircraft with the relevant type of engine (even a hard worked airline engine is unlikely to do more that about 3000 flight hours in a year and most will do much less).

The maximum acceptable probability for a **Catastrophic** failure leading to almost inevitable loss of the aircraft, is 10 times lower, 1 in 1000 million hours, which is not achievable with a practical single component and can, in general, only be achieved where more than one unrelated failure occurring simultaneously is necessary before the **Catastrophic** event could result.

THE AIRCRAFT METHOD OF DEALING WITH CRITICAL PARTS

The means of demonstrating an acceptable probability of **Hazardous** failure of **Critical Parts** before the aircraft goes into service, has been developed over many years by the major engine manufacturers and in general, the results of operation of the large numbers of engines in service show that they have got it right.

Using our turbine disc referred to above as an example, the basic method for establishing a satisfactory probability of failure for a single **Critical Part** depends on 5 simultaneous approaches.

1 Detailed stress and temperature analysis of the disc, covering all phases of engine operation in service and all anticipated atmospheric conditions to find where the highest stress levels occur and the stress levels at various engine operating conditions.

2 Extensive testing of individual sample discs mounted in spinning rigs with a large number of cycles of acceleration and deceleration to establish the fatigue life of the

disc, to which a large factor is applied to cover the full range of variations in material fatigue properties established by laboratory tests on a large number of material samples.

3 Very careful control of the production material, to ensure that the probability of material defects is minimised, and consistent control of the manufacturing processes and inspection of production discs to ensure that no sub-standard discs are supplied to the engines.

4 A mandatory maximum service life for each individual disc, with usage to be recorded against an individual disc serial number throughout the life if the disc. This maximum service life takes into account the demonstrated fatigue life of the disc and the intended use in service operation. For example, the anticipated operating conditions and time at take-off power, climb, cruise, descent and reverse thrust on landing. This is normally called the flight profile and the safe life of an engine component would be defined as a number of these flight cycles. A review of the actual flight cycle from service history would be carried out periodically to ensure that the originally defined flight profile was being followed and, if not, the **Critical Part** safe lives would be adjusted to maintain the same level of safety.

5 Maintenance of records which allow the detailed history of each individual disc to be traced back to the original material and manufacturing records so that, if any defect comes to light in service, other discs with potentially the same problem can be quickly found and withdrawn.

Sketch 2 below shows a simplified version of a typical turbine blade and disc assembly in an aircraft gas turbine engine to illustrate the different loads applied to the disc, all of which play a part in the overall fatigue life of the disc. These loads come from centrifugal force on blades and disc, gas loads on blades driving the turbine, pressure differences between front - rear of both disc and blades, thermal expansion loads due to temperature differences between rim and hub, drive flange loads transmitting drive to the compressor, and possibly high frequency vibration loads from the blades. The resulting stress patterns in the disc from the combination of loads are obviously very complicated and change with different engine operating conditions so that fatigue life usage per flight cycle is not easy.

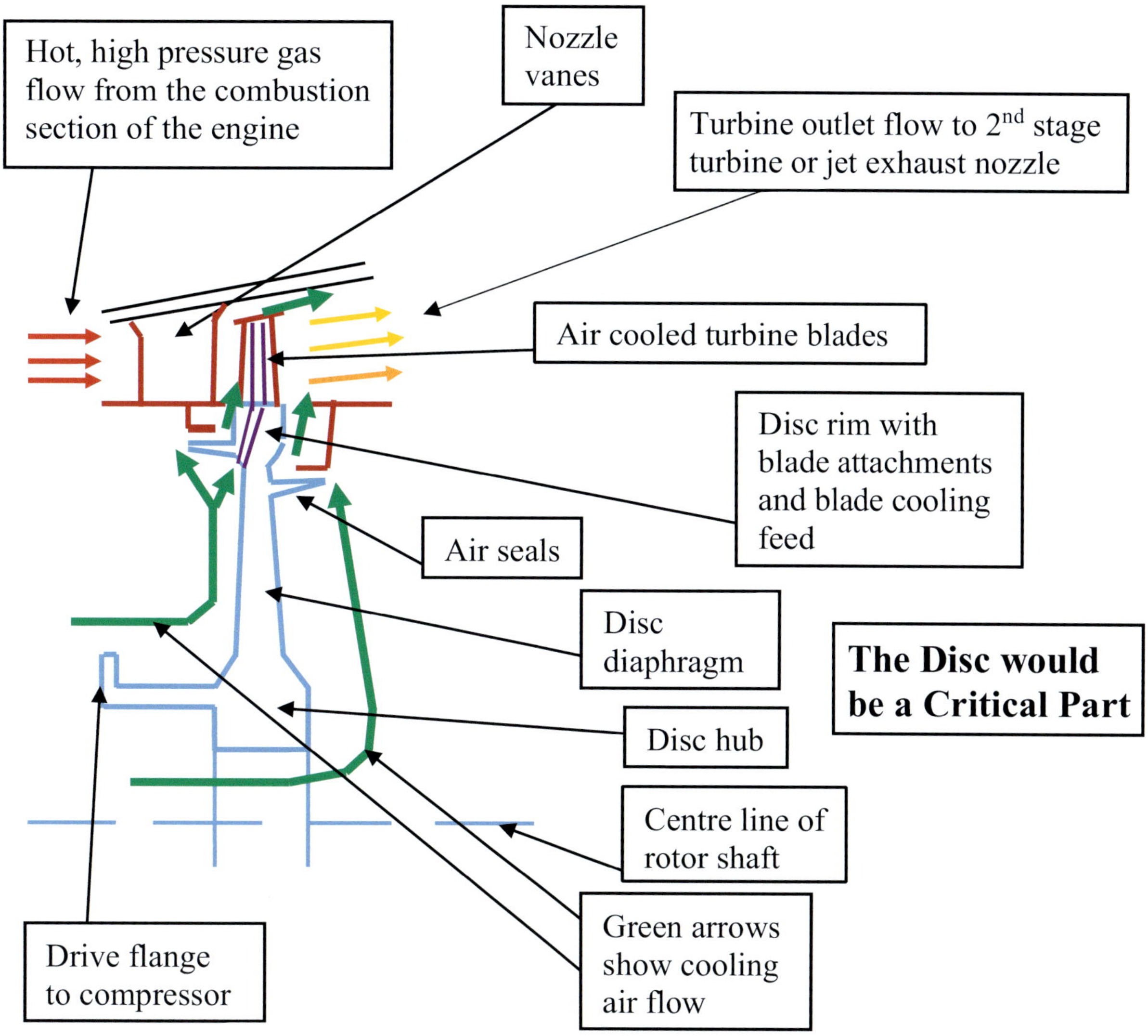

Sketch 2 – Typical aircraft engine turbine.

Non-rotating **Critical parts** are generally avoided as far as possible by provision of alternative load paths in case of any likely potential failure to avoid a **Hazard**. If a static **Critical** part is unavoidable, similar precautions to those described for our turbine disc would be necessary although it is sometimes possible, as weight control, although important, is somewhat less essential than on a rotating part, to give a static part so much extra reserve strength relative to associated parts that failure of the **Critical Part** is not reasonably expected to occur. Think of a steel solid engine mounting bracket, which could be a **Critical Part**, attached to the engine by means of numerous much weaker parts, bolts, links etc., such that it is virtually impossible for the bracket to fail before failure of one of the many attachments, thus allowing any deterioration to be found during routine maintenance before any potential **Hazard** could develop.

This introduces the concept of a mechanical fuse. In the same way as an electrical fuse protects a circuit from damage due to overload, sometimes a deliberately weaker part can be incorporated in a design in a position where it will fail first in any overload situation and by doing so protect a **Critical part** from overload. One extreme aero engine example of this type of design is sometimes used in the bearing supports for a very heavy, high-speed rotor such as a gas turbine fan. These fans are required, by certification rules, to be able to withstand the impact of a large bird or the loss of a complete fan blade without causing a **Hazard** to the aircraft. Damage of this type will always cause a high level of out of balance in the rotor which puts very high rotating radial loads on the supporting bearings. If a mechanical fuse is included in the bearing supports, an overload failure of the fuse will allow the bearing to orbit within suitable limits, allowing the rotor to run closer to its new centre of balance and reducing the out of balance loads.

There is not much scope for the use of mechanical fuses on a motorcycle but the general principle that, where possible, ensuring that the first part to fail in any overload situation does not result in an immediate **Hazardous** outcome is worth bearing in mind.

APPLICATION TO ONE-OFF MOTORCYCLES

As stated, the above is a lot of work so we need to consider what we can do on our one-off motorcycle to achieve a similar level of safety without spending a lot of time on the subject.

i) The first most obvious point is to know which parts are **Critical** in your design. Most are obvious and common to all bikes such as the fork legs, cast wheels, etc., but some are more obscure and require a bit of thought, for instance the gear selector forks where failure, in some cases, could result in two gears becoming engaged simultaneously, locking the back wheel.

ii) The second step is to review the design to see if the need for a **Critical Part** can be avoided by providing some fail-safe arrangements such as the bump stop of last resort previously mentioned in Chapter 3 to avoid a locked wheel in the case of failure of a mono-shock rear suspension unit mounting.

iii) The third step is to justify an acceptable probability of failure for the remaining **Critical Parts**.

My method of doing this is as follows:

Wherever possible, use parts from a production machine of equivalent or greater weight to make use of their control of manufacturing quality, their testing and the experience of satisfactory use over many miles on many machines on the road, and modifying the parts, if necessary, without reducing strength. In the absence of any other information, machine weight is a good guide to the loads experienced by almost any component as static loads and maximum bump loads are directly related to weight and, given sufficient power, acceleration, braking, and cornering forces are all limited by tyre adhesion which again is related to weight. With the large number of different motorcycles available and the vast range of parts, there is a standard part which can be modified to suit most requirements available somewhere, perhaps even from a car. It is just a case of finding it.

iv) The last resort is to make a one-off **Critical Part**, but that requires a bit of thought on strength, fatigue life and material and manufacturing quality.

One case I remember from my own experience is a good example of this and covers all the points in the 5 aspects of the aircraft process with nothing more than a bit of thought before you start.

Sometime during the 1970s I needed a front wheel spindle to attach a Honda 2 leading shoe racing drum brake to a leading front link fork arrangement for a one-off special (the bike was a supercharged 200 cc Velocette Valiant with a homemade centrifugal supercharger, an attempt to keep up with the small two strokes – see Photo 4). I was convinced of the advantages of leading link front suspension at the time (of which more later), and built two specials with the system, but it does require the design and manufacture of a number of "one-off" ***Critical Parts*** *so needs a bit of thought.*

Material was the first question and, having no easy access to suitable new material, I made the wheel spindle from a car front wheel drive shaft from a Mini.

This satisfied the material quality as there were large numbers of these shafts on the road and I had never heard of one breaking although maximum stress levels and fatigue conditions in the car, which was having success in rallying and racing, exceeded those anticipated for the front wheel spindle. I had also seen a severely bent one following an accident with no sign of cracking, providing good evidence of ductility.

The shaft was from my own car, so I knew its history. It had low mileage, and no accident damage and was replaced because of damage to the splines and circlip groove which had nothing to do with strength. The wheel spindle was machined from the drive shaft and carefully inspected, and crack tested before fitting. Thus, the first three points, stress analysis, sample testing and material and manufacturing quality, where adequately covered. The spindle was satisfactory in use, but I did take it out and inspect it a few times before I had complete confidence in it.

The other two points concerning monitoring in service were covered by this occasional inspection and the fact that total mileage and thus fatigue life usage would be low compared with the very large number of miles including racing and rallying demonstrated by the cars.

Thus, all the 5 main points in the aircraft method were adequately covered at very little cost; much better than using material of unknown origin or properties, or a piece of an unsuitable grade of soft stainless steel because of its impressive shiny appearance.

There are no existing photographs of the Velocette spindle but Photo 2 below shows another example of a similar homemade ***Critical part****, in this case a stub axle for an on-going 3-wheeler project. This spindle was made from new material, a high-quality steel of medium tensile strength and good ductility. Dimensions were dictated by the existing wheel bearing and hub carrier dimensions but resulted in high strength reserves using the pessimistic assumptions for maximum loads as described in Chapter 4, section 1.*

Out of interest, Photo 3 shows the spindle assembled in the hub carrier, another ***Critical Part****, but in this case borrowed from a car, and positioned in the part assembled front suspension. I am using Austin Metro hub carriers fitted upside down and on the opposite side of the machine. In this position they are suitably dished and have the top and bottom steering swivel ball joints it the right positions to allow a motorcycle front wheel to be fitted and still maintain an almost centre point intersection between the steering axis and the tyre centre line.*

There is nearly always some proven part for any ***Critical*** *position if you can find it.*

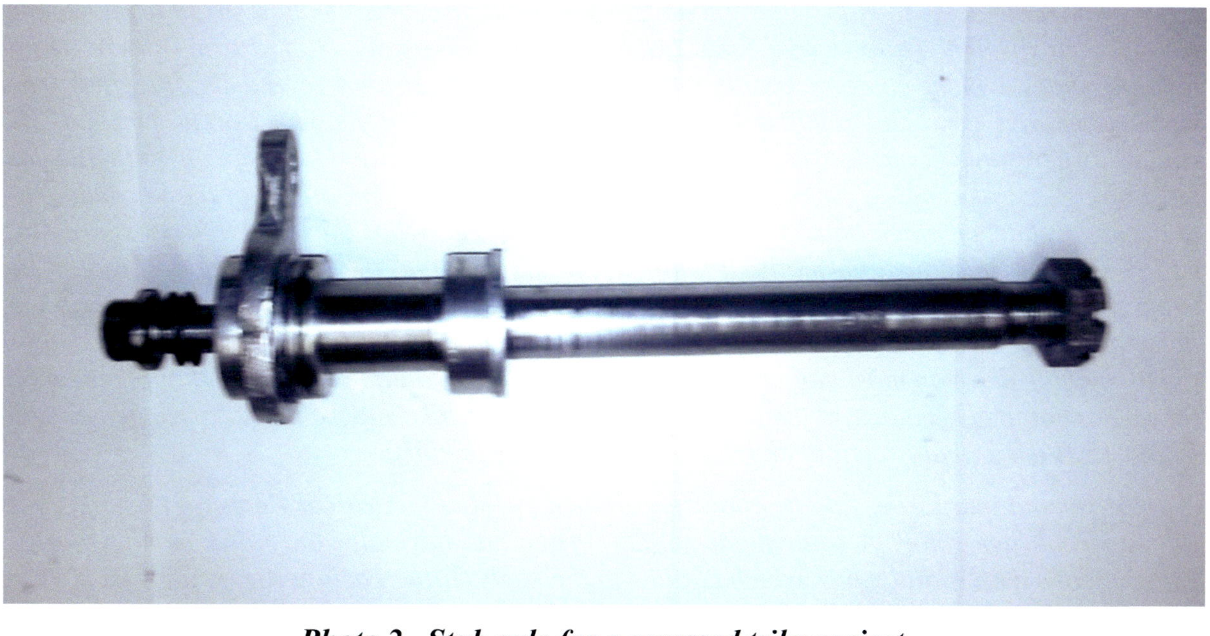

Photo 2 –Stub axle for a-reversed trike project.
(Part 5 includes more details of the trike, now completed.)

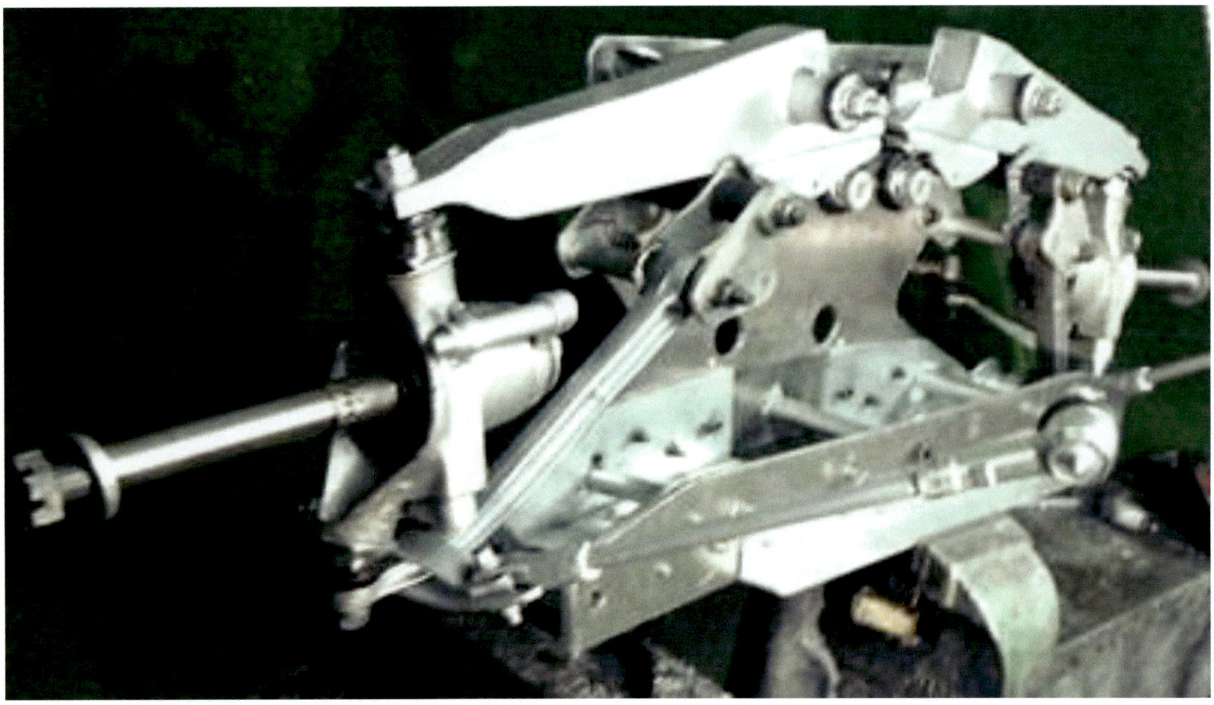

Photo 3 –The stub axles assembled into reversed car hub carriers and fitted to the part assembled trike front suspension. *The only other homemade Critical parts are the top suspension arms, some parts of the supporting plates and two small brackets (not shown) which hold the front ends of the bottom arm forward tie rods. The remainder of the Critical parts are all taken from other vehicles where they are subjected to higher loads than they will be on the trike and modified as necessary without reducing strength. The top arms, support plates and tie rod brackets are designed with large strength reserves by the method described in chapter 4. Motorcycle rear suspension units will be used, angled outwards from the ball joint shown below the inner end of the top arms and the lower outer corners of the support plates.*

*Back to the supercharged Velocette, after being distracted again into thinking about the homemade **Critical** parts for a more recent project.*

The modified Valiant, flat twin, a similar design in principle to a miniature boxer twin BMW, was a nice bike to ride with lightweight, good handling and the best drum front braking I had ever experienced up to that time. I used the large diameter Honda drum brake with 2 leading shoes which, combined with 80% anti-dive incorporated in the leading link front suspension gave good feel combined with enough power to lock the wheel on a dry road. The supercharger gave some top end power boost but never enough to keep up with the fast developing 2 strokes of the period. One of the problems was the limitations of the Vee belt drive to the supercharger which limited the supercharger speed and thus the available boost pressure (if the belt was tensioned normally, it slipped and if very tight, the belt tended to break). The supercharger really needed to run at twice engine speed, but this proved to be impossible to achieve reliably with the Vee belt drive system. A modern thin Multi Vee belt drive as used for the alternator drive on most modern cars or a toothed belt would have solved that problem but they were not in general use at the time. Photo 4 shows the bike supercharger arrangement and Photo 13 (Part 2 photo section) shows other features of the bike.

*A second interesting failure analysis problem, related to **Critical Parts** on this bike, was the supercharger rotor. A centrifugal supercharger has basically only one moving part, a rotating disc with radial vanes which rotates in a closely fitting casing. Air or air-fuel mixture enters at the centre of the disc and flows outwards between the vanes due to centrifugal effects. As the gas moves outwards its speed is increased by the disc vanes until it emerges at the outer rim at high speed. The kinetic energy of the high speed gas flow is then converted into boost* *pressure by means of a diffuser, a long shallow angle conical duct where the gas speed reduces gradually as the area of the duct increases. My rotor was an aluminium sand casting, incorporating both the disc and the drive shaft, produced by a local foundry from a homemade pattern and finished by machining and hand work. The failure analysis problem results from the large amount of energy stored in the disc due to its high speed of rotation, such that if it were to break up, perhaps due to fatigue cracking, the pieces of disc would fly off in all directions — not the sort of thing you would want to happen between your legs at speed. With the uncontrolled methods of manufacture, it was impossible to absolutely guarantee the integrity of the disc, so the possibility of failure had to be considered. The answer was to make the outer parts of the casing strong enough to keep any detached parts of the rotor inside the casing, thus avoiding the* ***Hazard*** *and the need to make the rotor a* ***Critical Part****. The debris containment capability added about 2 lb of weight to the cast aluminium casing but removed one worry about the safety of the bike. Photo 4 below shows the supercharger mounted above the engine with a belt drive from the front of the crankshaft (the constant vacuum SU carburettor shown at the supercharger inlet supplied the fuel mixture but had no throttle butterfly. The throttle slides, provided by 2 Amal carburettors with no fuel feed, were downstream close to the cylinder heads).*

Much more efficient centrifugal compressors are available today, as used in car turbo chargers although, being of smaller diameter, they need to run at much higher speed. If I was to attempt a similar modification again, I would probably use one of these, the high speed mechanical drive would then be the only problem to be solved. Using an exhaust driven turbocharger avoids the mechanical drive problem but has never been very successful on bikes because of the need for absolutely predictable and instantaneous control of engine power. Some lag in power build-up after throttle movement and subsequent power increase at constant throttle is acceptable in cars but could be very unsettling on a bike on a slippery road.

It seems that mechanically driven superchargers are making a comeback these days, particularly on small cars, so perhaps a small capacity supercharged engine could again be a good option for a special.

Photo 4 – The Velocette supercharger.

SECONDARY EFFECTS ON CRITICAL PARTS

Unfortunately, that is not the whole story in ensuring the safety of **Critical Parts**.

A satisfactory probability of "primary failure" of a **Critical Part**, that is failure of the part while operating in the intended design operating conditions with no other defects in the system, can be established by the above methods, but if there is some other failure which may damage or affect the operating conditions of the **Critical Part**, the carefully demonstrated low probability may not help.

Back to our aircraft example of a turbine disc. Maximum safe disc life is based on material properties and also on operating temperatures and speed of rotation. Increased temperature results in reduced material properties and increased speed of rotation increases stress levels, both of which would reduce safe life or, in the extreme case, could cause an immediate **Hazardous** failure.

Thus, obvious threats to the satisfactory probability of failure are over-speed above the design limit or over-temperature relative to normal operation. The other possibilities are rubbing contact with static parts generating extra heat and reducing metal thickness, or corrosion which would accelerate the development of metal fatigue.

In general, in an aero engine the maximum speed of rotation will be controlled by a system of limiters, usually electronic these days, but mechanical on earlier engines (using the principle of rotating bob weights which originated in early steam engines). These limiters work by reducing fuel flow to reduce power and prevent any impending over-speed, while maintaining the pre-set limit speed.

Over-speed is thus normally prevented by the limiter but if there was a failure of the limiter or the fuel control system, an over-speed would still be possible. There is also the possibility of sudden failure of the turbine drive shaft which would leave the very high power developed by the turbine driving nothing, resulting in very rapid turbine acceleration, possibly to burst speed before the limiter could operate.

In order to maintain the satisfactory probability of turbine disc failure, in spite of this additional over-speed risk, some sort of additional system is required and this is provided by a totally independent over-speed protection system, normally an electronic system these days, set to operate from independent turbine speed sensors at a speed just above the normal speed limit.

The over-speed protection system would normally totally shut off the fuel supply very rapidly and thus shut the engine down. This type of system is known as a "Safety System" and plays no part in normal operation.

This introduces yet another problem as, if a Safety System plays no part in normal operation but needs to work in a small fraction of a second on the one occasion, perhaps in the whole life of the engine, when it is required, how do we know if the Safety System has itself become inoperative over years of lying dormant? With a mechanical system we can only rely on mandatory periodic inspection, but an electronic system has the advantage here as it can be made to be self-checking and thus draw attention to any faults which may have developed for repair during maintenance.

Over-temperature is the other main threat to the integrity of our turbine disc and could result from a reduction in the disc cooling airflow or a temperature increase in the cooling air, or more likely from an internal fire as a result of oil leakage into the cooling air. The temperatures are so high in the turbine area of a gas turbine that even the cooling air is hot enough for self-ignition of any oil leaking into it and the resulting oil fire could overheat the turbine discs. The usual answer to this threat is to continuously measure the cooling air outlet temperature and use this to provide a warning of high cooling air temperature before the turbine disc overheats significantly.

The possibility of rubbing of the disc on a static part is very low as failure analysis of the design would have ensured that all likely rub positions are eliminated during the design process. Corrosion is usually covered these days by the use of material which does not corrode significantly or by the use of the best available sacrificial coatings.

The above notes on the possibility of secondary effects of over-speed or over-temperature may be interesting but perhaps not very relevant to motorcycles. The most likely cases on a motorcycle would result from either corrosion of a **Critical part** or rubbing contact between rotating and static parts.

Safety systems are sometimes useful in protecting potential critical parts, a common motoring example being the safety chain, which is required in the UK when connecting a trailer to a towing vehicle, to guard against the tow ball joint becoming disconnected. The chain plays no part in normal towing but would prevent the trailer running loose out of control if the main towing connection was lost.

Corrosion of **Critical parts** to the point of failure is not often seen on motorcycles but is a real concern on older cars. Something like a badly rusted girder fork or steel tube frame on a vintage machine for restoration would be a bit suspect in this respect. In particular, any open-ended tube is always suspect as they are often not well protected on the inside and internal corrosion cannot easily be seem. I always blank off any open-ended tube, after putting a bit of "Waxoil" or a similar product inside, to prevent water ingress.

The extreme example of the risk of a corrosion induced **Critical part** failure is a magnesium wheel or frame part. Not much magnesium is used on motorcycles due to cost and manufacturing difficulties and its potential for deep intergranular corrosion with little to see on the surface before the part fails. It has a great strength to weight advantage over aluminium or steel for some types of part and may therefore be tempting in attempts to produce a machine with ultra-low weight.

There is also a phenomenon called stress corrosion where an initial small fatigue crack originating from an initial corrosion pit is accelerated by further corrosion within the crack until the part fails as a result of fatigue, cracking at a much lower life than would have been anticipated. The only answer to this one is, if you must use magnesium or any other material prone to rapid corrosion for a **Critical Part**, to make sure you protect it with the best available coating and inspect it often.

An example of possible contact between rotating and static parts on a motorcycle could perhaps be contact with some minor part. A mudguard support, or some other static part close to the wheel, could become displaced and rub on the spokes of a wire spoke wheel which could, in the worst case, eventually cause the spokes to fail and the wheel to collapse.

GETTING THINGS WRONG

The most important principle in all things related to engineering is:

"Do it once and do it right."

Unfortunately, occasionally we all make mistakes in the making or assembly of parts, typically taking too much metal off or holes in the wrong place, or stripped threads during assembly, but when a mistake is made with a **Critical** part, it is often too expensive or time consuming to scrap it and start again. The main thing not to do in these circumstances is to cover up the problem and carry on regardless, but instead to think through the possible effect of the defect on the probability of a **Hazardous** failure. There is usually a safe way out of the problem, but you will not find it until you have considered what the effects of doing nothing would be.

The aircraft system of dealing with this sort of problem is based on formal systems which are always based on four principles.

i) Avoid a culture of blame, which does not fix the problem and could discourage reporting of errors.

ii) A technical review of the possible effects of the problem by the appropriate engineering people.

iii) A decision, fully recorded and signed off, which could be "safe for unrestricted use", "repairable", "restricted use", or in the worst case, "scrap."

iv) Permanent marking of the part with a number or symbol to allow details of the problem and the engineering decision on usage to be traced through carefully maintained records.

This system, rigorously applied, will allow no deviation from the intended design to get into flight without full engineering approval.

Occasional mistakes are almost inevitable, and I have made my fair share over the years but two, where I was at fault, have stuck in my mind, both a long time ago when I was working much closer to the metal than in later years.

The first was during my apprentice days when we spent some time working with the aircraft fitters assembling the centre fuselage section of the Canberra aircraft. A large number of pipes and cables ran along the roof of the lower bay under the fuselage and were clipped to the fuselage ribs by means of pipe clips and small bolts, and the job was to drill holes for the bolts to dimensions on a drawing. The centre fuselage was a flat ended, more or less parallel, cylinder and, unless you were totally familiar with all the detail, it could be difficult to tell the front from the back. Unfortunately, I got confused and drilled a hole in a rib in exactly the correct position on the wrong side of the aircraft. This could have been a big deal as the ribs, when assembled, are an integral part of the structure and cannot be removed and replaced. A company system, similar to that described above, kicked in and it was concluded that the hole was in a low stress area and had no effect on the integrity of the aircraft, so the deviation was recorded, signed off and the hole filled with a rivet. Not the best way to find out about the method, but part of my education.

These days this type of hole would be accurately positioned and drilled during manufacture of the rib so that no drilling on the assembled part would be necessary.

There is another old engineering saying, common in the machine shops at the time:

"Measure twice and cut once."

Meaning be very sure you are cutting in the right place before cutting metal.

I understand that a Canberra which at one time held the world altitude record is currently being restored to flying order by the Vulcan in the Skies Trust.

It is very unlikely to be the same aircraft as a large number were made, but if they find a 3/16 hole, filled with a rivet, in one of the centre fuselage frames, which is not on the drawings and serves no useful purpose, they can blame me.

The "mirror image error" is very easy to make and I did it again quite recently on a part for one of my projects, so it is one to beware of when you are making something which is almost, but not quite, reversible.

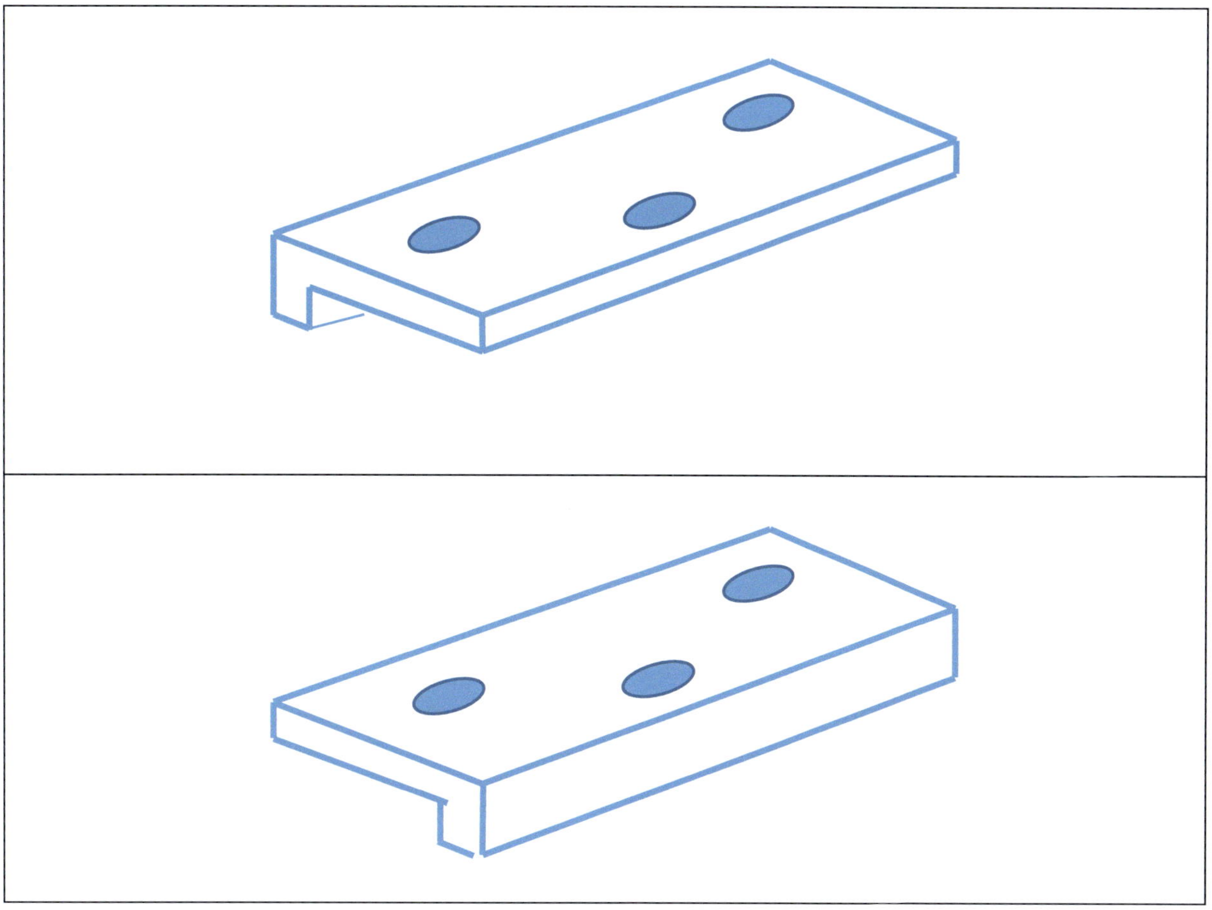

Sketch 3 – A typical mirror image error.
The pattern of holes is the same but could easily be drilled with the part the wrong way round.

The second remembered error was a well-known trap which I fell into and should have known better. It is usual to assemble a gas turbine with the engine stood on one end with the shafts vertical, as it makes it easier to ensure concentricity of the heavy rotors with no weight effects pulling them to the lower side. I was leaning over the top of an assembled engine looking at someone else's problem when a pen fell out of my top pocket and disappeared down between the blades. There was no need for a technical review in this case, as the pen had to be extracted or it could have damaged the blades when the engine was rotated, and dismantling to reach the pen would have been a very time consuming and costly process.

We turned the engine upside down on a crane and rocked it about and fortunately the pen fell out, so all was well, but I was not too popular for a while. A similar problem on a bike would be to drop a small nut down a spark plug hole, a problem which cannot be ignored and is easy to avoid but is very inconvenient to fix after it has happened.

Chapter 4 – Quality Control

The causes of a large majority of the small number of potentially **Hazardous** failures which do occur on aircraft have some element of human error, for example, less than ideal material or manufacturing quality, or failures in the strict adherence to maintenance and operating procedures. The aircraft certification process, including the failure analysis, ensures that there is a very low probability of a primary failure of a **Critical Part** which is of the specified quality and is operated and maintained correctly, but it is obvious that, if the part or its material are sub-standard or maintenance procedures are not followed, the predicted low probability of failure which justifies safe operation is invalid.

I am sure that the same applies to the manufacture of the large majority of standard motorcycle and motor car **Critical Parts**, particularly in these days of type approval and potential legal liabilities, but unlike the aircraft manufacturer, the road vehicle manufacturer has no control of the use or abuse of the finished vehicle.

The low probability of failure of **Critical Parts** achieved on mass production road vehicles is therefore probably as a result of lower stress levels (see chapter 5) relative to material strength, that is, greater strength reserves than would be used in aircraft design where minimising weight is essential. This would allow greater margins of safety for the abuse which may occur in use and a fatigue life which could exceed the life of the vehicle.

For this reason, if a suitable standard part can be found for our motorcycle **Critical Part** requirements, or one which can be modified to fit without reducing its strength, it is obviously a safer option than designing and making something new, where ensuring that the quality of the finished part is entirely your own responsibility.

It is also worth considering that proprietary custom parts and parts designed for racing may have lower strength reserves than standard road-going parts in the search for weight reduction. Also, some may not have all the necessary controls on material quality or manufacturing processes and will certainly not have anything like the testing or road-going mileage of production parts. If they are to be used in **Critical** positions on our one-off motorcycle, a bit of extra thought is needed to ensure that they will be safe and durable for the intended use.

If a one-off **Critical Part** to your own design cannot be avoided, something like the following approach will ensure that all the above points are covered:

1) Make the best possible estimate of the loads applied to the part.

Very detailed analysis methods are available for determining stress distribution throughout a component these days, which ensure optimum designs for sufficient strength with minimum weight, but this sort of detailed analysis depends on accurate knowledge of the maximum loads to be applied to the part and thus is not usually available to the special builder even if he has the capability to do the analysis.

*In the absence of any better information I have used 10 times the static loads, excluding road bump loads etc. — that is the maximum loads imposed by weight, acceleration, braking and cornering, which are easy to calculate, as the design limit load for **Critical Parts**, taking no account of the (difficult to define) loads due to road surface irregularities or other undefined loads. Then design for maximum stress levels within the proof strength of the material at these conditions (see definitions, chapter 5). This provides a large margin to cover extreme transient loads such as landing after jumps, mounting curbs, potholes and other road bump loads etc., and will also give a good fatigue life as the large majority of use will be with much lower loads than the above factored design load.*

*Although this approximate method results in parts which are undoubtedly somewhat heavier than the optimised part would be, they have a large margin of safety which is essential on **Critical Parts**. The resulting parts also tend to have similar dimensions to typical production parts for the same type of use which may or may not be a coincidence.*

I offer no guarantee for this method but have never had any problems with parts designed in this way.

2) Do not use unknown material or used material with unknown history for a Critical Part.

Good quality material with a known specification is half the battle, the rest being careful accurate machining etc. and some sort of detailed inspection of the finished part for minor cracking or material defects. A good inspection usually can be carried out with just a magnifying glass and a good light, paying particular attention to points where there may be stress concentrations. If in doubt, there are other methods, e.g. dye penetrant, X-ray, ultrasonic or magnetic crack detection, but these would normally have to be done by a company with the right equipment.

It is worth noting that polishing to a mirror finish does nothing to improve fatigue life and may be detrimental as it can prevent detection of minor cracks. The polishing can disguise a crack by burring the surface metal over the crack. The moral is if you must polish the **Critical Part** for cosmetic purposes, do the crack inspection before you polish.

The same argument applies to electroplating and shot or sand blasting, although shot blasting tends to improve fatigue life slightly by introducing compressive stress in the surface of the material. Electroplating can have the opposite effect, tending to reduce fatigue life on some materials due to a phenomenon called hydrogen embitterment.

*Another diversion from the subject, but related to the design of **Critical Parts**. I recently saw a modern 750 cc motorcycle which had been in a front-end accident. One of the fork legs had sheared off below the bottom yoke with the other on the point of a similar failure. The fork tubes were of marginally larger diameter than traditional classic bike forks but the tube walls where remarkably thin when compared with a typical classic bike. I assume that a high strength steel had been used to save weight, but the thin wall section had the unfortunate effect of allowing the tube to buckle into an oval cross section under the yoke when overloaded in the accident. The tubes had then spit along the sides of the oval leaving two shallow "U" sections with almost no strength in bending. A classic fork leg in similar conditions would probably have just bent back without buckling or breaking. Not much help in an accident but it just illustrates the many factors affected by choice of material for a **Critical Part**. For interest, Sketch 4 below shows the effect.*

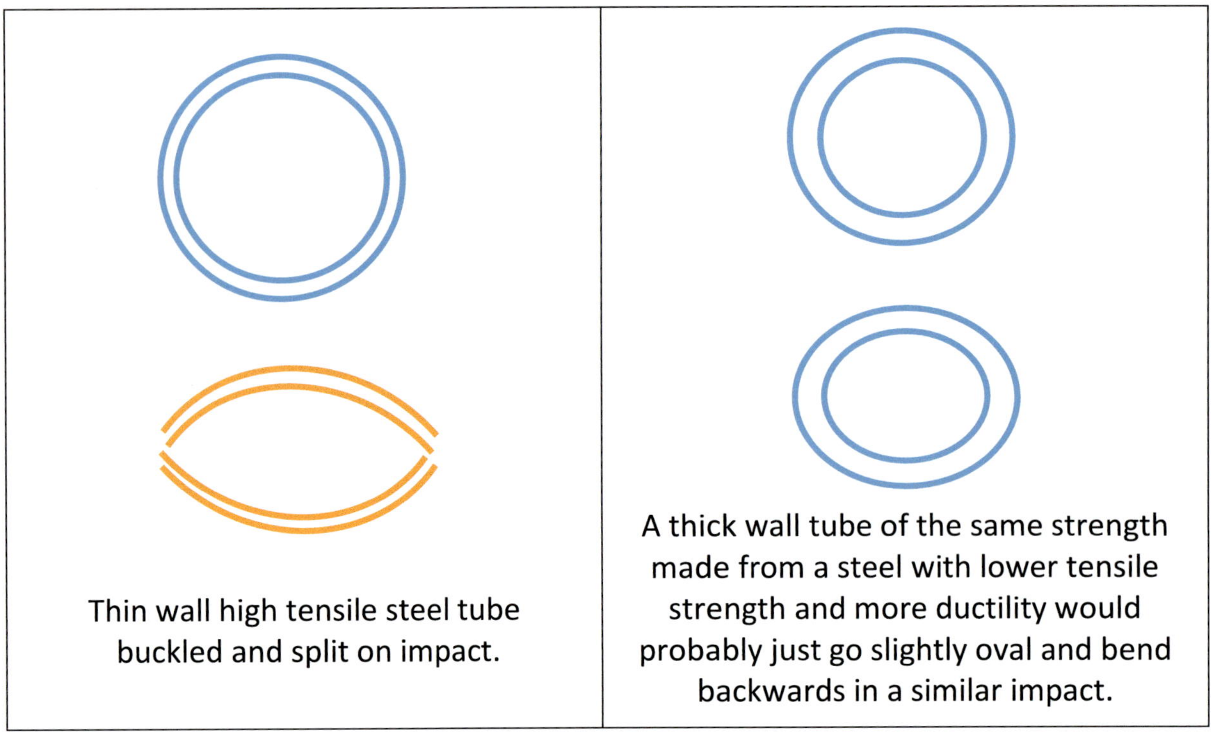

Sketch 4 – Two tubes with the same strength in bending but very different failures when overloaded in an impact situation.

Another memory prompted by the second sketch above was an unfortunate incident with a cow when travelling to work on my 350 cc AJS in the early morning sometime in the 1950s. At that time, I had a daily commute of almost 40 miles each way, mostly on what we would now call back roads, the only roads in my area at that time. The Preston bypass, the first UK motorway, was opened around that time but was no help as it crossed my route at 90°. I was travelling quite slowly at the time but impact with the wandering cow in the middle of the road bent the forks back about 2 inches although I stayed upright. The cow just got up and walked away but the fork tubes were bent below the bottom yoke exactly as shown in the second sketch. They cost about £12 to replace, a big sum in those days. Tough, solid animals, cows.

The AJS, which was never modified, was totally reliable on this daily commute in all weathers and returned over 60 mpg. I wonder if a modern commuter with the same power (about 16 bhp) from a smaller capacity would do the job as well.

Chapter 5 – Definitions

I apologise to those readers who are motorcycle enthusiasts but do not claim to be engineers for drifting into the use of some engineering terms in the previous chapters, so I have included this section in the hope of making all clear. Engineers may prefer to skip the chapter as an exercise in stating the obvious, but these things are only obvious if you are familiar with them.

Loads – The loads are the forces applied to the part in use together with the position at which these forces are applied and the direction of the forces.

Stress – This is the load divided by the cross-sectional area of the part supporting the load.

Thus, stress is measured in terms of load per unit area and can be compared with material strength as quoted in the material specification in the same terms.

Stress can be **Tensile** (stretching the part), or **Compressive** (squeezing the part). There are also **Shear** stresses where the applied load is parallel to, but not aligned with, the supporting feature as shown in Sketch 5.

Tensile and compressive stresses are self-explanatory, but shear may not be as clear. A motorcycle example of shear loads would be the gas pressure and inertia loads which are transmitted from an engine piston to the connecting rod through the gudgeon pin. These loads are transmitted across the two planes at either side of the connecting rod small end bearing by means of a shear stress on the pin.

When the applied loads are bending the part, stress levels vary through the cross-section of the part but are highest at the outer surfaces with one side in tension and the other in compression.

A compressive load on a long part of relatively small cross-section is likely to result in buckling at much lower load than would cause a failure in compression. Buckling is a failure where the part in compression fails in bending in a direction at 90° to the line of the loads. It occurs because any small deviation from an absolutely straight part with load applied exactly along the centre line will cause the compression load to be distributed unevenly over the cross-section and therefore will also apply a small bending load. This bending load will result in a small amount of curvature in the part which will increase the deviation from absolute straightness and further increase the bending load and the curvature. It is easy to see that the greater the deviation from a straight part becomes, the greater the effect of the compressive load on increasing bending until the part fails in bending.

Torsion, (twisting a part), introduces shear stresses in the material, for instance in a torsion bar type spring, such as a car anti-roll bar, or a coil spring which is really just a torsion bar wrapped round into a spiral. A torsion bar also has the highest shear stresses on the outer surface, but in this case they are equal all round the bar.

Sketch 5 below illustrates the different types of load which could be applied to our part and the resulting stress in the part.

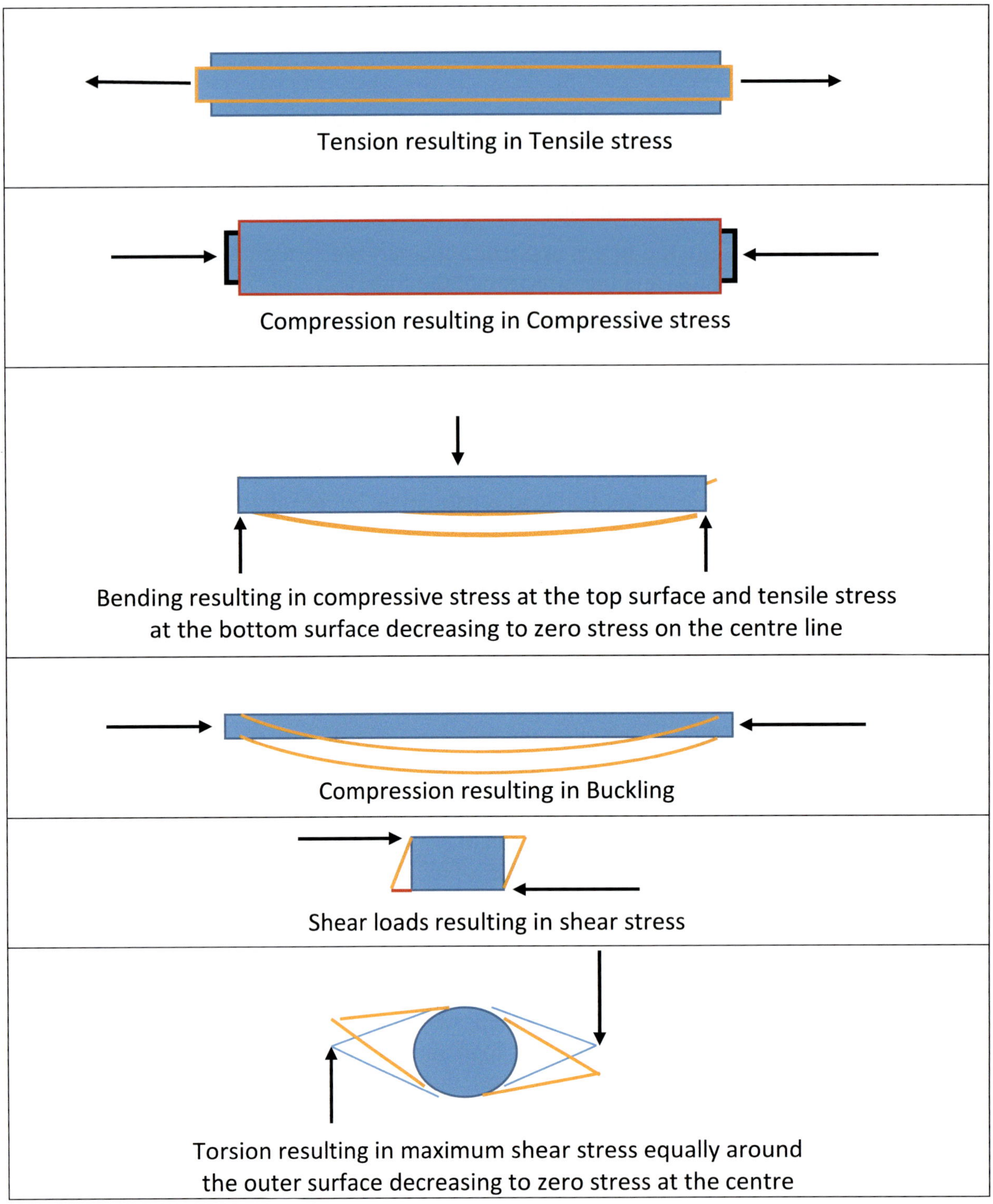

Sketch 5 – Types of load and resulting stress in a part.
(The red lines indicate the effect of the load.)

Limit loads. These are the maximum loads which may be applied to the part in normal use. For example, loads resulting from maximum weight, maximum engine power, maximum tyre adhesion and maximum bump loads from road irregularities.

(As maximum bump loads are difficult to determine, the approximate method described on chapter 4 above excludes bump loads from the **design limit load** and covers them with a large reserve factor).

Some people will call me old-fashioned, but I always use the imperial system (stresses in pounds/square inch (lb/in^2 or psi) for motorcycle calculations as it is much easier to visualise *than the international standard SI metric system, which our politicians tried to force on us in the 1970s, perhaps to demonstrate their European credentials. The SI system uses the pascal as the basic unit of pressure or stress (a very small unit, 6895 pascals is equal to one lb/in^2) or kilopascals (kpa) and megapascals (mpa), (equivalent to 1000 and 1,000,000 pascals, that is 0.145 lb/in^2 and 145.0 lb/in^2 respectively).*

These units are not so easy to automatically visualise as imperial units, and thus make it very easy to lose a factor of 10 in the calculation without realising it.

The basic problem with the system is that the standard SI unit of length, the metre, is far too large relative to the standard unit of force, the newton (0.225 lbs force); 1 newton is the force needed to accelerate a 1 kilogram weight to a speed of 1 metre per second in one second.

One pascal is the stress or pressure resulting from the very small force of one newton distributed over a very large area, one square metre, and as a result is a very small stress or pressure.

It is all totally logical but also totally inconvenient to use. Must have been originated by academics rather than practical engineers.

For those who totally disagree with me on this, I always ask them some question on the lines of "If I said 690 kpa to you, would that be likely to be a typical tyre pressure, a workshop air supply pressure or a hydraulic system pressure?" Answer immediately without mentally converting to imperial and you are a fully metric engineer. It is surprising how few people, including engineers, can immediately visualise the metric figure in the UK after over 50 years of exposure to the system, although it may be different in other countries.

In spite of all political efforts by people who use units for nothing more complicated than buying a pound of potatoes or a litre of petrol, the imperial system survives, not least in the USA where they adopt a more sensible policy of using whatever unit is most suitable for the job in hand.

In one way, I benefited from the conversion in that I was able to obtain a complete set of high quality imperial measuring instruments, micrometers, verniers, high gauges, slip gauges etc. for a knock down price when the company machine shop converted to metric. These are still in use today and are as accurate as when they were new with no electronics to fail or go out of calibration.

Enough of the politics — to continue with definitions.

Material strength. Material strength is usually defined as a Yield Strength which is the maximum stress which can be supported without resulting in any permanent stretching of the material after the load is removed. As the exact yield point is difficult to find, yield strength is usually specified as a **Proof Strength** where a small permanent stretch of 0.1 or 0.2% of original length occurs. This is the material strength we should use for our design calculations.

Ultimate Strength is also specified and is the maximum stress before breakage occurs. Stresses between the proof and ultimate strength of a material will result in permanent deformation of our part when the load is removed; that is to say it will be permanently bent or stretched.

The proof and ultimate strength of materials normally used in motorcycle modification can easily be found from reference sources providing that you know the material specification reference number. If modifying an existing part for another use and the exact material specification is not known, the safe method would be to check that the maximum stress in your use of the part is not higher than that in the original use, or to use the lowest listed material proof strength for the class of materials, e.g. stainless steels, aluminium alloys etc.

Reserve factor. The reserve factor shows the proportion of the material strength not used when the maximum (limit) load is applied to the part. A very large reserve factor will therefore result in a heavy part which will never break and a reserve factor of one would mean the lightest possible part which would be just on the point of yield at maximum load.

Very low reserves are used on some aircraft parts, particularly in non-critical applications, in pursuit of lightweight but only after very detailed analysis of loads and stress levels and only for very occasional load cases where fatigue is not a factor.

For our motorcycle we need relatively large reserves to cover uncertainty in the maximum loads, particularly bump loads, and to ensure a satisfactory fatigue life.

Fatigue. Fatigue is a failure mechanism where repeated application of loads much lower than the design limit load can cause progressive development of very small initial cracks or material defects in a part until the part fails because the remaining un-cracked material is insufficient to support the load.

Fatigue usually starts at points of stress concentration or initial material defects resulting from manufacturing or from corrosion, wear or damage marks or machining score marks. Cracks originating from this point then penetrate further into the material with every subsequent load cycle.

Fatigue life is the number of load applications required to fail the part and in general needs to be greater than the expected life of the machine.

The rate of propagation of a fatigue crack is very variable even in the same part of identical material subjected to identical loads, so it is usual to put a large factor of safety, (typically 6) on any single fatigue test result to establish a safe life for the part. (This factor can be reduced if several identical tests on different specimens of the part are carried out.)

Stress Concentration. In a straight rod under tension, the stress level is equal across the whole cross-sectional area of the part but for more complex shapes and loading conditions stress levels will vary through the part. Any internal corner, thin section or hole in the part tends to introduce a stress concentration where the local stress level will be higher than the general level, sometimes two or three times higher. The effect of a small crack or corrosion pit at a stress concentration will increase the effect and it is easy to see why these positions are ideal sites for fatigue to start.

A good example of the effect of stress concentrations on fatigue life is our clutch plate problems on a Yamaha TZ350 two-stroke. A friend of mine had the bike and was racing it on the short circuits but was having endless problems with cracked clutch plates, cracking from the corners of the outer dogs. Whether the rider was exceptionally hard on the clutch, or the plates were sub-standard, perhaps pattern parts rather than original equipment, I never discovered but it was costing a lot of time and money. The plates appeared to have been made by a milling or perhaps a stamping process which had resulted in a very sharp corner at the inner end of each dog with burred edges on one side — an ideal site for initiation of fatigue cracking at a point where there would be a large concentration of stress as the drive load from the outer dogs is transmitted into the body of the clutch plate. We took a new set of plates and put a radius (0.050 inches approximately) in the corner of each dog with a Swiss file to both reduce the stress concentration and remove the sharp corner and burred edges. The job was a bit time consuming but worth it in time and money saved by greatly increasing the life of the plates. Taking metal off can sometimes make the part stronger if you remove it on the right places.

Another example of avoiding unnecessary stress concentrations, this time by design comes to mind from a car modification I did a long time ago. We needed two drive shafts to connect a Rover differential unit to Lotus wheel hubs on a rear wheel drive car. The shafts also acted as the top link of an independent suspension system and therefore were ***Critical parts*** *as* *failure would have allowed the wheel move out of position with only the bottom link to retain it, not good for staying in control of the direction of travel.*

Unfortunately, the standard shafts which fitted the differential were tubular, approximately 2 inches (52mm) diameter and the shafts which fitted the wheels were solid, approximately 1.25 inches (32mm) diameter. As no shafts which fitted both ends were readily available, the obvious solution was to make composite shafts from the inner end of one and the outer end of the other. The first thought was to make a bush to suit the inner diameter of the tube and the outer diameter of the smaller shaft, assemble to the right length, as shown on Sketch 6, and then weld the bush to both shafts, at points A and B (Sketch 6).

This solution would probably have been satisfactory, but you cannot be too careful with a ***Critical*** *part. "Probably," in my opinion, is not really good enough when failure could send you straight through the nearest hedge or worse. The snag was that the highest stress in the shaft, a combination of torsion from transmitting drive to the wheel and tension or compression from side loads on the wheel, would be at the outer diameter of the smaller shaft, not the best place to have an uncontrolled weld which could introduce stress concentrations and therefore reduce fatigue life.*

The eventual solution was to make the bush a very tight interference fit with both shafts so that the welds were not really needed to transmit the drive, then fit the bush to the smaller shaft and weld at the low stress-free end, point C as shown in Sketch 6. Then press the small shaft and bush into the tubular shaft and weld on the low stress outside end, point A.

It needed a large hydraulic press to assemble the shafts, but fortunately a local garage proprietor, whose hobby was restoring vintage Albion wagons, for which he had collected a variety of heavy equipment, was very helpful in this. There is nothing like having the right contacts when needed.

The finished shafts had satisfactory straightness and balance with no reduction in strength or fatigue life. A much better solution for a ***Critical*** *part, from a failure analysis point of view, than the first thoughts.*

Reducing stress concentrations in a clutch plate

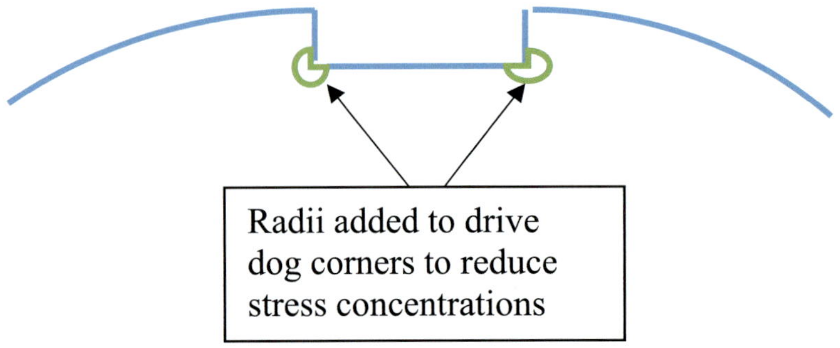

This small change increased the fatigue life of the plates enough to effectively eliminate the cracking problem

Drive shaft - Stress concentration avoided by design

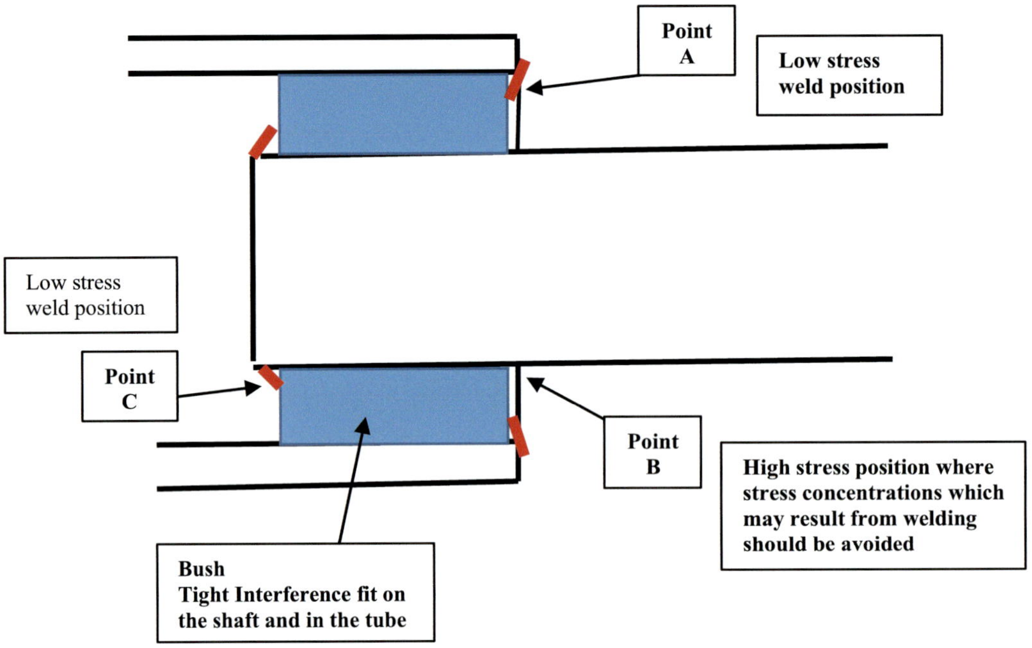

Sketch 6 – Examples of reduction and avoidance of stress concentrations.

RESIDUAL STRESS

When a part is made involving cold bending, stretching, welding etc., stresses are put on the material during the process. Some of these stresses can become trapped in the part with areas of compression and areas of tension which balance each other internally with no external loads applied.

These internal stresses are extremely difficult to detect in the finished part but have the effect of reducing the material strength available to carry external loads if, for instance, they add further tension to an area of the part already under residual tension.

Relief of residual stresses can usually be achieved by heating the whole part to high temperature and cooling it slowly and evenly, as is often done with aircraft engine **Critical** parts such as forged turbine discs, but this is not usually practical for unique **Critical** parts made for a motorcycle modification.

As a simple example to show how residual stress can occur, think of a large, thick steel plate with a large hole in the middle. Then attach a thin, straight steel rod across the hole by welding it to the plate at both ends. The rod will be heated much more than the plate during welding, which will only be at welding temperature in a small area close to the weld, and will expand, increasing its length much more than the plate.

After the welds have solidified, the part will cool but the rod will be prevented from contracting to its original cold length as it is now attached to the plate at both ends.

The rod will then be in held permanent tension by the much stronger plate, that is residual stress will be trapped in the rod.

The problem could have been easily eliminated by design if the rod had been bent into a shallow "S" shape before welding, allowing the contraction during cooling to partially straighten the rod as it contracted to its cold length.

The only answer for the special builder is to be aware of the possibility of residual stresses when deciding how to make a part and, if it is unavoidable, allow sufficient strength reserves on any part where high residual stress is likely to result from a lot of bending or stretching of the material or welding being used in making the part.

PROBABILITY

In aircraft failure analysis, the probability of each possible **Hazardous** failure is quantified in numerical terms for the whole fleet of aircraft of a particular type with limits for the maximum acceptable number of failures per million hours of flight operations. There is a maximum acceptable probability for each category of failure, i.e. **Minor, Major, Hazardous** or **Catastrophic**, which must be demonstrated by test or theoretical analysis or by comparison with similar parts which are satisfactory in service operation.

There are two different types of failure where probability can be a factor in deciding what is acceptable.

The first, which may be called random failures, is when a small proportion of identical parts fail prematurely but the majority have no problems. An example of this would be a material defect or an unsatisfactory weld causing a failure where good material or good weld would be safe. It is not possible to put a numerical probability of this type of failure directly before a number of failures have occurred so, if the weld is in a **Critical** position, extreme care must be taken to ensure that all examples of the part are of the highest possible material and manufacturing quality. Aircraft and engine manufacturers will have collected large amounts of data on failure rates on existing components made from the same material by the same manufacturing methods and so will be able to use this to justify an acceptable numerical probability.

The second type is a failure which results from long periods of use and will occur eventually on every example of the part. This may be as a result of fatigue or possibly wear or corrosion or some other reason, but in every case there will be a safe life after which the probability of failure will rise rapidly with further use. If the resulting failure could be **Hazardous**, the only safe solution is to routinely replace the parts with new before they reach their safe life limit.

It is not really possible to apply the same methods to a one-off motorcycle, but we need to achieve the same effect in each case where a single part failure would result in a **Hazard**. That is that the probability of a **Hazardous** failure in the likely total usage of our machine is sufficiently low that we can accept the risk.

You could consider that one cannot totally eliminate the possibility of being struck by lightning, but the risk is so low that it is not worth worrying about. This is the effect we are aiming for with each possible individual **Hazard** on our motorcycle.

I suggest that we should always aim for at least the same level of safety as an equivalent production motorcycle which is in general use, for each aspect of the machine, whether it is a one-off part, a proprietary part or one borrowed from some production motorcycle. Then we can refer to the satisfactory usage of the large number of production machines over a much greater mileage to justify a satisfactory probability of failure for each aspect of our machine.

The unfortunate thing about probabilities is that they add up. When there are many possible causes of a **Hazardous** failure, each one may be acceptably improbable but the probability that any one of the many causes will result in a **Hazard** occurring is much greater than that from any single cause. For this reason, we should look at all possible causes, no matter how unlikely, to try to eliminate as many as possible and to make the total probability of any one of the remainder occurring as low as possible.

Chapter 6 – Causes of Failure

In order to look at the failure analysis of a part, we first need to consider the possible causes of failure which may affect the part. In general, for metal mechanical or structural parts, there are 5 main causes of failure: Overload, Overheating, Fatigue, Wear or Corrosion, but not all of these will be applicable to every part.

Non-metallic parts such as plastics and composite parts, for example glass fibre or carbon fibre are in general subject to the same possible causes of failure but the failures tend to develop in different ways. As most of our **Critical parts** will be metal, these notes are aimed mainly at metal parts. The peculiarities of **Critical parts** made of non-metallic materials such as carbon fibre wheels and frames are referred to in the sections of part 3 covering frames and wheels.

OVERLOAD

This is failure as a result of a single application of a much higher load than we have designed for, and in general will not happen if we have got the design right except in cases of the use of a sub-standard material, which is not as strong as the intended material, or as a result of an accident.

Thus, providing we are sure of our design and material quality, the probability of a failure of our motorcycle parts as a result of overload in any "non-accident" situation is very low.

From a failure investigation point of view, a typical overload failure will generally show signs of stretching of the material before failure, for example a rod in tension would show a reduction in diameter tapering towards the failure point, known as necking. The failure surface would usually be an irregular series of peaks and troughs often in the form of cones and reverse cones. Complete failure of a metal part in direct compression is very unlikely but, for a long relatively thin part, such as a push rod, the most likely type of failure in compression due to overload would be buckling (see Chapter 5 Sketch 5). If a direct compression overload failure did occur, the effect would depend on the ductility of the material. A fairly ductile material such as mild steel or soft aluminium would just compress in the direction of the load with the displaced (squeezed) material spreading out sideways into something like a barrel shape. A more brittle material such as cast iron or some magnesium alloys would tend to fail in shear in a plane at an angle of approximately 45° to the direction of the load, with one half of the part slipping alongside the other.

One example of overload failure I remember from long ago in the 1950s was a Royal Enfield trials bike which was dropped when riding through a deep stream. It drew in a full cylinder of incompressible water at fairly high revs and the inertia of the heavy flywheels was enough to lift the cylinder head off the barrel, stretch the cylinder base studs and bend the connecting rod (buckling due to overload in compression). At least it proved that the engine design was good with equivalent strength in all parts. A strong part attached to weaker parts is only as strong as the weakest link.

OVERHEATING

Engine overheating due to loss of cooling water etc. will generally only result in loss of power. If this does not alert the rider to the problem, the engine may eventually seize but even this will generally be manageable with the clutch and thus not be **Hazardous**.

The effect of more severe overheating to very high temperatures is to reduce the strength of materials and if this reduction is sufficient to exceed the strength reserves, the part may fail as a result of "overload" at loads lower than the normal design load. The sort of temperatures we are talking about vary from material to material and can be found in material data sheets if you know the exact material specification. Unfortunately, in most cases a special builder will not know the exact specification and in that case, it is a fair assumption that anything over about 200°C for aluminium parts or over 300 to 400°C for steel is likely to result in a significant reduction in strength, although some specialist steel alloys retain their strength at much higher temperatures.

Very few motorcycle parts are at risk of causing a **Hazardous** situation as a result of severe overheating. The exhaust valve, piston and exhaust system may be overheated as a result of unsatisfactory fuelling, ignition timing or cooling or the brakes could overheat due to rubbing, but otherwise there are no sources of sufficient heat for significant overheating to occur. The possibility of fire, which could provide sufficient heat, is covered in the section on fuel systems in part 3 of this book.

Therefore, failure of a **Critical part** of our motorcycle as a result of overheating, with the possible exception of the brakes, is unlikely but the possibility should still be considered in failure analysis in order to rule it out in most cases.

Overheating is a much more serious problem in a gas turbine aero engine because of the generally very high temperatures in the combustion and turbine sections of the engine where turbine blades, for example, can be running in gas temperatures above the melting point of the turbine blade material and the turbine discs and perhaps other "hot end" parts will be **Critical parts**. These parts are totally reliant on cooling air flowing around the discs and through passages within the blades to keep the metal temperatures low enough to avoid overheating failures, and the cooling air itself can be hot enough to cause self-ignition of any fuel or oil leaking into it (Sketch 2 in Chapter 3 shows a typical turbine arrangement). The high cooling air temperatures result from the necessity to compress the cooling air to a pressure higher than the gas pressure in the turbine to ensure that it will flow into the turbine against the surrounding pressure. A proportion of the work done in compressing the cooling air is transformed into heat in the same way as the air compressed in a diesel engine cylinder is raised to a temperature high enough to ignite the fuel and cooling air pressures in a modern aero gas turbine could be higher than the compression pressure in a diesel.

FATIGUE

Fatigue, as described in chapter 5, is the most likely cause of failure for any motorcycle **Critical part** as it can occur in normal operation after a sufficient period of repeated operating loads less than the maximum design load. The probability of fatigue failure of our **Critical parts** can be minimised by:

- Sufficient strength reserves in the design
- Good quality material
- Design to minimise stress concentrations
- Radiusing of sharp edges and internal sharp corners
- Careful inspection for surface defects scores or corrosion particularly at positions where there are stress concentrations
- Protection against corrosion

As fatigue cracks generally develop fairly slowly, there is a good chance of finding them before the part fails, if you bother to look. This is a very good reason for keeping the bike clean. There is no better way of ensuring that every exposed part is looked at regularly than cleaning it.

After they have happened, which could be a bit late in the case of a **Critical** part, a fatigue failure is usually easy to identify by looking at the failure surface with a magnifying glass.

There are usually a series of small ridges and troughs, known as striations, surrounding the point where the fatigue started, known as the origin, which is normally at a point of stress concentration or a surface defect. The striations then show the progress of the developing crack. The diagram included in Sketch 7 shows the general idea. With greater magnification it is often possible to estimate the number of load cycles to failure by counting striations.

Fatigue cracking can result from a relatively small number of high loads such as may result from off road riding at speed over a very rough terrain, normally known as low cycle fatigue, or from a very large number of smaller loads from engine vibration or a similar cause, known as high cycle fatigue.

I have not always followed my own advice in respect of regular cleaning and remember one case on my modified Vincent Rapide (Photos 5, 6 and 11) when on holiday in the Isle of Man at TT week in the early 1960s. At that time the bike was in relatively standard form, before many of the modifications shown in the photos, so the problem was not related to the modifications. The cast aluminium rear wheel hub failed at the flange behind the brake drum and spoke flange, as shown in Sketch 7 below, as a result of fatigue cracking which must have been developing for a long time but was hidden covered in dirt and grease. The IOM bumps, much worse in those days, finished it off. When we eventually got the bike home (in the days before insurance breakdown cover was common) we made a better hub, machined from solid using a stronger aircraft aluminium alloy with thicker flanges, larger radii and with the sharp edges at the ends of all the holes radiused. That hub is probably still going somewhere today although I sold the bike in 1972 for £72 (should have kept it as it would have been worth a fortune today).

As I remember it, the inner end of the bolts securing the spoke flange and brake drum to the hub were almost impossible to reach and hold for tightening of the nuts and were designed with eccentric cylindrical heads intended to prevent rotation during tightening. Vincent design was usually based on sound engineering and much better than most manufacturers at the time, but they must have had a bad day when deciding on this type of anti-rotation feature. The problem with it is that tightening torque applies a very high outward load to the bolt hole and a very high local compressive stress to the hub, as shown in the sketch, which is further increased if the hole becomes slightly oversized. The hub flanges transmit all the loads between the wheel and the bike and were therefore subject to a large number of fatigue load cycles during running.

It is likely that the origin of the fatigue was at the point marked X in the sketch, and it developed from this point around the flange.

On the replacement hub the eccentric head bolts were replaced with socket heads, unusual on bikes at the time but common these days. These could be held for tightening with a long hexagonal key and did not require an anti-rotation feature.

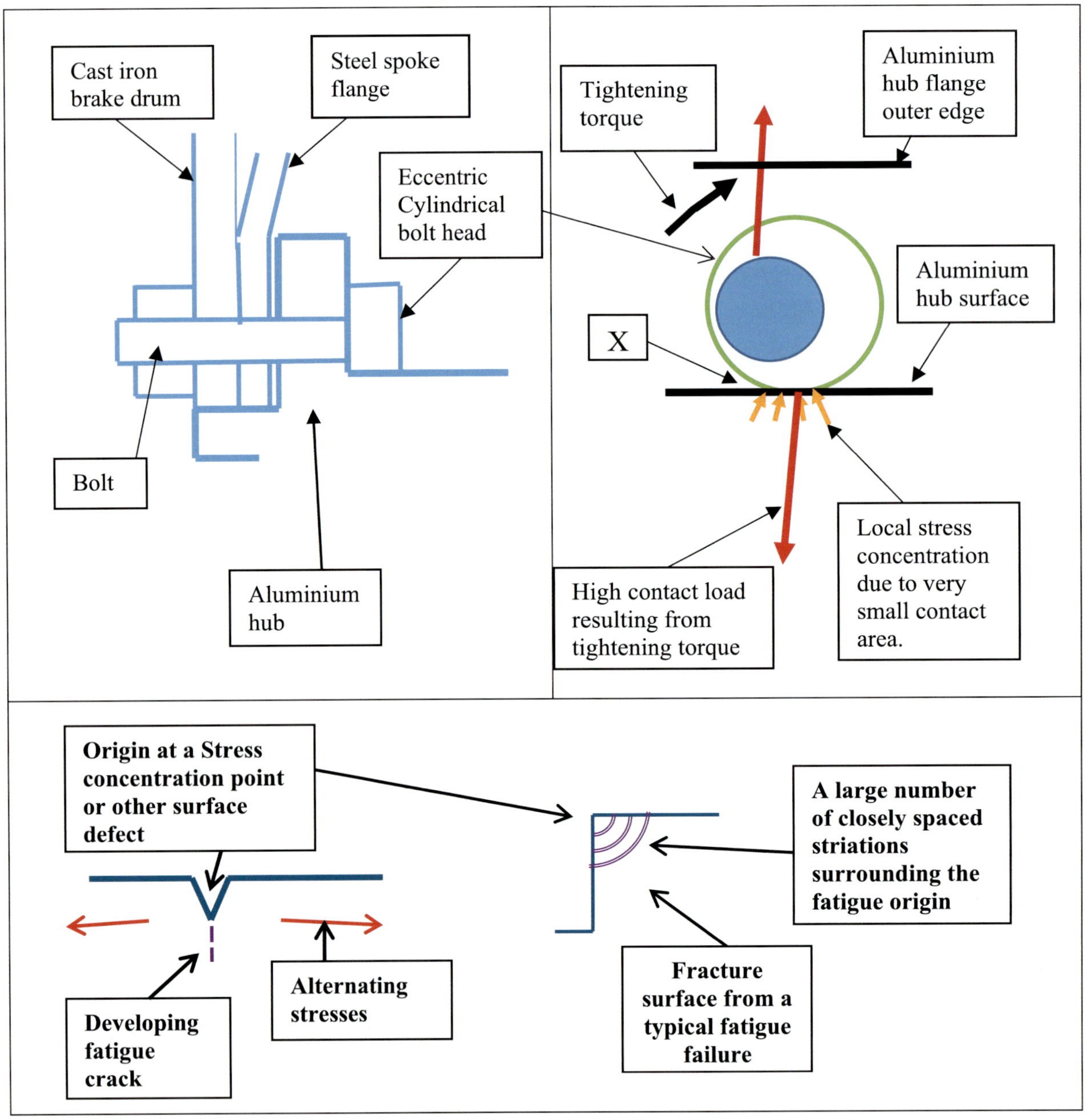

Sketch 7 – Section through one of the flange bolts and bolt anti- rotation feature (the bottom sketch shows the method of recognising a fatigue failure from the fracture surface).

I hope Vincent enthusiasts will excuse any minor errors in the Vincent sketches, drawn from memory from a long time ago, but they illustrate the point that very small design features can result in major failures if they result in stress concentrations which initiate fatigue cracking. In this case, the contact load resulting from tightening the bolt would be dependent on how much of the tightening torque was absorbed in friction in the thread and at the nut face, but could easily be several 100 lbs of load on a very small area line contact. The aluminium would then locally deform to give a small depression with just enough contact area to support the load, resulting in a local ultimate compressive stress surrounded by an area of high local tensile stress.

Photo 5 below was taken during later modifications to the rear suspension, but shows how congested the rear hub area was with the two brake drums, the reason why the hub bolt anti- rotation feature was necessary.

I subsequently removed the brake on this side, just keeping the single brake on the other side. This change reduced un-sprung weight which had no effect on rear braking. With rod or cable operation, dual brakes do not improve braking power as, with the same lever lengths, they only distribute the same pedal load over a bigger friction contact area. The only real gain is in fade resistance and friction lining life which are not very important on a back brake. Hydraulic operation, as with disc brakes, is different in that two brakes will double the braking power at the same lever pressure and the same lever arrangement and master cylinder, the difference being that with the hydraulic operation the lever travel would also double.

Photo 5 – The Vincent during modification to the rear suspension showing the congested rear hub area
which makes the hub bolts inaccessible. The finished modifications are shown in Photo 6.

Photo 6 – The Vincent in its final form.

This photo shows the Vincent in its final form with a single rear brake and after all the other modifications. It never got a clutch cover as the Vincent one would not fit over the Norton clutch. The open clutch gave no problems and there is a lot to be said for the quotation from the USA:

"If it ain't broke don't fix it."

I learned a lot from this bike, mainly that enthusiasm and confidence are no substitute for sound engineering and judgement of risk.

The notes on Photo 11 list all the main modifications.

Another diversion off the subject, but related to fatigue. Around the same time as I had the Vincent, I was working on a long-forgotten aircraft called the TSR2 which was cancelled during its flight test period for political and financial reasons rather than engineering reasons. I have been involved in many and various fatigue problems on engines and aircraft over the years, but this one has stuck in my memory, probably because it was the first time I really became involved with the subject, previously having worked mainly on fuel and hydraulic systems.

This aircraft was intended to be able to fly long distances at high speed at very low altitude to avoid detection and had two large engines which were much less fuel efficient that today's engines. For these reasons it needed to carry a lot of fuel and every spare internal space was used as a fuel tank including the rear fuselage. A section of fuselage behind the wing was used as a tank and surrounded the two jet pipes which connected the engines to the jet exhaust nozzles at the rear. From a failure analysis point of view, this was not the best place to have a large tank of fuel; next to two jet pipes which operated at very high temperatures, well above the spontaneous ignition temperature of the fuel, but the safety case was to be covered by filling the "air space" above the fuel in the tank with inert gas to exclude oxygen and thus prevent fuel ignition or generation of an explosive mixture of fuel vapour and air in the tank. This left the question of possible fuel leakage from the tank into the space surrounding the jet pipes being ignited by contact with the hot jet pipes. A high-speed flow of air through this space was intended to carry away any leaking fuel out of the back of the aircraft before significant fire damage could occur. The maximum speed at which a flame can spread through a fuel air mixture can be measured and, as long as the airspeed is greater than this, no fire can stabilise at any point in the airstream; the same principle as blowing out a candle.

We were given the job of doing a cyclic fatigue test on this tank to demonstrate fatigue life and also effective fuel sealing under service conditions. The tank was subjected to bending and torsion loads as part of the aircraft structure together with fuel filling and emptying and large variations in internal pressure and temperature, so a test rig was needed to simulate all these conditions on a static test on the ground.

Illustrating the point that first ideas are rarely the best way of doing any job, all sorts of weird and wonderful schemes involving numerous individually controlled hydraulic jacks and large roller bearings where proposed, none of which achieved exactly the required structural loading conditions of combined bending and torsion. Eventually one of the older draftsmen in the office, who was not working on the job, had a quick look and very quickly produced a sketch of a system which achieved the required test conditions exactly using only two jacks and a number of fixed length links. The arrangement seems obvious now, but at the time, somebody needed to propose it and once it came to general attention, all the other proposals were immediately abandoned. Sketch 8 shows the arrangement.

Simultaneously with the structural loading cycles, the test also involved repeated fuel filling and emptying cycles with hot and cold fuel, internal pressure cycles and heating cycles on the tunnel skins simulating heat from the jet pipes. The outer skin of the tank would be exposed to outside air temperature which could be as low as -50°C at high altitude resulting in additional stresses due to differential expansion between the inner and outer skins, particularly when the tank was empty and the fuel did not provide a heat sink.

In the days before programmed electronic controls, the test was controlled by a mixture of manual controls and simple electrical systems and, these days, would have been a health and safety man's nightmare (standing well back when the more hazardous parts of the test were in progress was the best advice available). These days, this sort of fatigue testing would probably be done with water or some non-flammable liquid in place of fuel for Health and Safety reasons, with the fuel sealing to avoid leaks accessed separately. Not as realistic but a lot safer.

The test went on for months, running day and night without many problems. The rig was generally reliable and the test tank survived with no significant cracking or leakage and subsequently gave no problems in the aircraft.

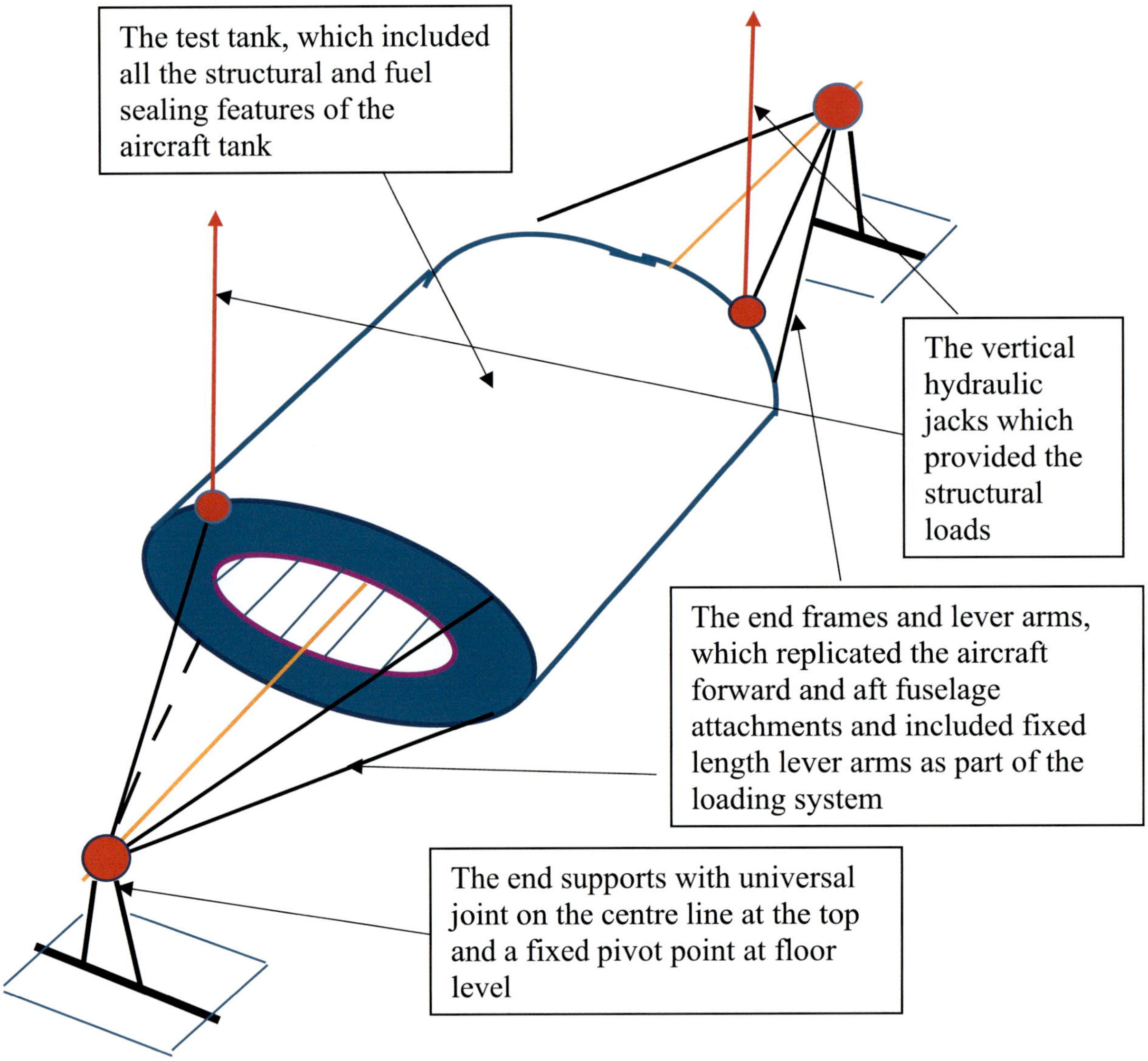

Sketch 8 – The fatigue test rig.

Equal vertical loads on the two hydraulic jacks gives constant bending movement and constant torque along the whole length of the test tank. The length of the fore and aft lever arms and the offset of the hydraulic jacks from the vertical centre line controlled the ratio of bending movement to torque. A sequence of varying hydraulic loads was then used to simulate the fatigue loads and stress levels which would be applied to the tank in flight.

The other thing I remember about this test is that it was the only time when I worked a regular sequence of day and night shifts; not to be recommended, but in those days it was a way of life for many people in the industry.

The economics of this type of cancellation escapes me. To cancel a project after spending all the money required to design the aircraft, do a lot of the necessary engineering work and build prototypes for flight test and then to immediately, within a few days, cut up all the expensive hardware up and send everything for scrap, leaves you with nothing to show for the money spent. The requirement for the aircraft is still there so eventually you will have to spend even more money to obtain a replacement.

Perhaps the political requirement that you must never be proved to have made a wrong decision overrides everything else. A special builder short of funds would just park the project, leaving the option to go back to it later.

Two aircraft, by some miracle, survived the wholesale destruction. The second prototype, which was scheduled for its first flight on the day of the cancellation, is now in the RAF Museum at Cosford and the 4th prototype, which was nearing completion, is in the Imperial War Museum at Duxford. Worth a look if only to show the rapid pace of aircraft development in the 1950s and early 60s, bearing in mind that the Spitfire and Lancaster had been RAF front line aircraft only twenty years before.

History tends to repeat itself: as recently as 2010, the improved Nimrod maritime surveillance aircraft was cancelled for financial reasons during the flight test phase. By some chance I had also worked on this aircraft, helping with certification of the new engines, but our part of the job was finished and the flight test engines had been delivered and proved satisfactory in flight long before the cancellation.

Enough of all these memories from long ago — back to the subject.

Still on the subject of fatigue, vibration from any cause can result in fatigue damage, as it causes rapidly alternating stresses in the affected parts and therefore rapidly builds up a large number of stress cycles.

There are basically three types of vibration which could affect our motorcycle: engine vibration, vibration originating at the tyre contact point with the road, and aerodynamic vibration.

Some engine vibration is inevitable on any engine with reciprocating parts, although it is much reduced on engines with balance shafts. This engine vibration is transmitted through the engine mounting points to the frame, resulting in alternating stresses in the frame around the mounting point. In general, if the main cycle parts are designed with sufficient reserve strength, as suggested in chapter 4, additional fatigue damage as a result of engine vibration will not be significant but if felt vibration becomes uncomfortable for the rider, it may be possible to introduce rubber bushes at the engine mount positions or to change the crankshaft balance factor to direct a bigger proportion of engine out of balance vibration in a fore and aft direction and reduce the vertical component which is the cause of most felt vibration. It is largely a case of experimenting but either or both of these changes may improve the situation. The parts more likely to be affected are minor lightweight parts such as mirrors and fairing panels, where their natural frequency of vibration may correspond to the frequency of engine vibration, but they are unlikely to fail in a **Hazardous** manner.

Road surface induced vibration could result from, for instance, a road surface with a series of ridges, too closely spaced for the suspension to react fast enough to smooth the ride, would produce a vertical vibration through the steering head but, as with engine vibration, if the main cycle parts are designed with sufficient reserve strength, as suggested in chapter 4, additional fatigue damage as a result of this type of vibration should not significantly increase the risk of a **Hazardous** failure, but it may excite the natural vibration frequencies of lightweight parts.

There have been exceptions to this in extreme circumstances such as the desert racing in the USA, where frame breakage, due to fatigue or overload, or a combination of the two, is not unknown, and where higher strength reserves and more efforts to elimination of stress concentrations may be required.

Aerodynamic vibration is most likely to affect fairing panels or similar parts and is caused by the pressure of the air flow over the panel at speed causing the panel to move or change shape slightly such that the new shape reverses the air load on the panel thus reversing the movement. If this sequence of event corresponds to the natural vibration frequency of the panel, a vibration can build up to the point where the panel or its mountings could fail in fatigue (think of blowing on the edge of a piece of paper — if the conditions are just right, the paper will flutter backwards and forwards across the direction of the airflow). The only likely **Hazard** from this type of failure is if a failed panel could interfere with the steering. Nothing disastrous is likely to happen very quickly, so the best way to cover this possibility is to wait for some road testing and, if any significant vibration is found, it is usually easy to change the natural frequency of the panel by an addition of some stiffening or an additional mounting point.

WEAR

Wear can occur where parts can rub against each other, particularly if they are unlubricated. It is generally a slow process but eventually it may initiate a fatigue crack or remove sufficient metal to reduce the strength of the part with the risk of an overload failure. The answer is obviously to prevent parts rubbing together in an uncontrolled way by ensuring sufficient clearances between parts which are not solidly bolted together and clipping all pipes, cables etc. so that they cannot contact other parts.

Wear in bearings and other parts which are intended to be in contact, for example brake pads, is a different story but for a failure analysis point of view, as always, we must first decide what would happen if wear occurred and then think about the worst effects of likely wear and finally the probability of it happening.

I cannot think of a case where wear on a **Critical part** is likely to develop to a point where it resulted in a **Hazardous** failure on a motorcycle. An aircraft gas turbine example would be if some static part became displaced and rubbed on a main shaft connecting the compressor and turbine. If sufficient metal was removed by wear from the shaft and sufficient heat generated by friction, the shaft could fail resulting in over-speed of the released turbine which would, instantaneously, still be generating several thousand horsepower with nothing but its own inertia to drive. In this case, the failure analysis would have to rely on whatever safety system was fitted to prevent a **Hazardous** turbine disc burst as described in chapter 3.

*Old cars are much more likely to suffer wear of **Critical parts**, particularly in the suspension and steering, and in this respect the MOT test has had a beneficial effect. I remember one case, long before MOTs, I was driving a side valve Ford 100E Prefect when a track rod end dropped off as a result of wear of the ball until it could escape from the cup. The released wheel immediately went onto full lock and we survived with full opposite lock on the other wheel. It was a borrowed car, and that is my excuse for letting it get into that state, but they were different times when this type of thing was accepted as one of the hazards of life rather than an unacceptable breach of health and safety rules.*

On the subject of worn ball joints, it may not be just a 1950s problem long forgotten, although improvements in materials and particularly in sealing against water and grit have reduced the probability of severe wear. I was looking at a modern low volume sports car recently and noticed that the bottom link of the front suspension was connected to the upright by means of a vertical axis Rose type ball joint on the end of the lower suspension link. The ball joint was attached to the upright by means of a single nut below the ball joint. The suspension unit acted on this bottom link of the suspension such that the weight of the car was being transmitted through that ball joint and, if the joint became worn to the extent that the ball could pull through the axially loaded upper part of the cup, it seemed that the suspension may collapse. I did not check whether there was an independent bump stop but I hope there was. Sketch 9 illustrates the problem.

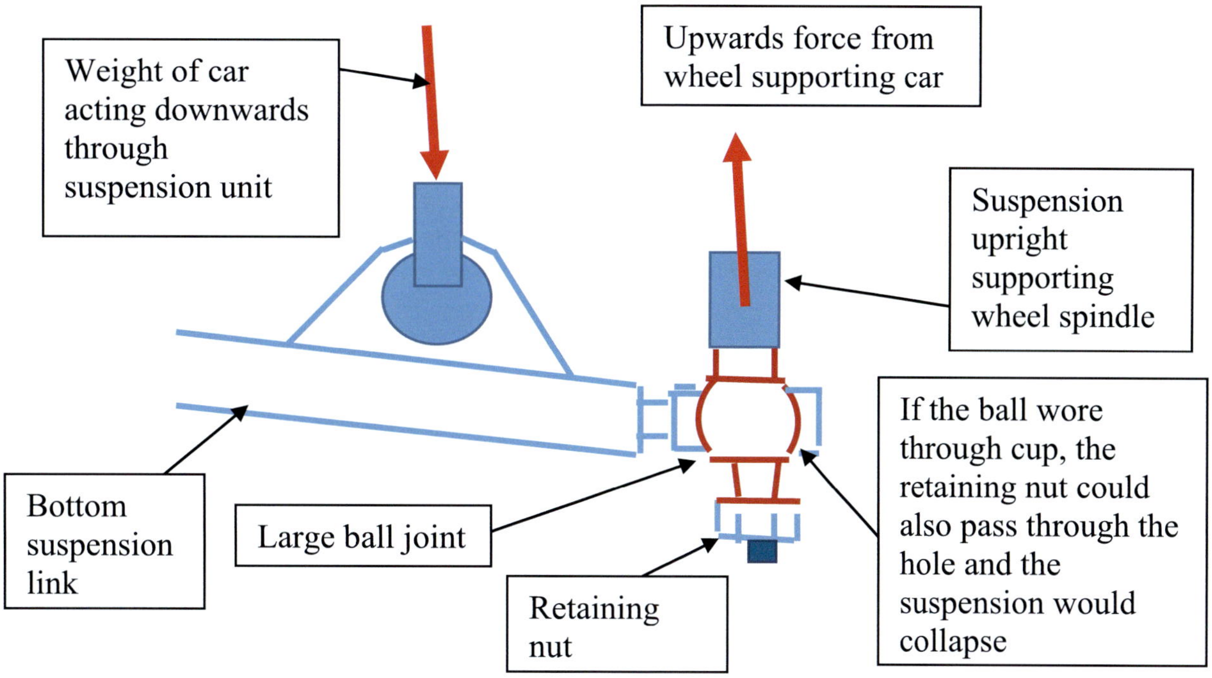

Sketch 9 – The car front suspension detail referred to above. A large washer between the retaining nut and the ball joint would avoid a potentially Hazardous failure.

No doubt the first owners of these cars will all be enthusiasts and will keep them in perfect condition, but all cars get old, and some get neglected. The obvious failure analysis solution would have been to make the nut outside diameter large enough or to add a large washer so that in the unlikely case of the ball wearing to the extent that it could pull out of the cup, the weight of the car would rest on the nut or washer and would prevent collapse until the resulting deterioration in handling drew attention to the failure. A potential ***Hazard****, however unlikely, which could have been eliminated at negligible cost.*

This example may not seem very relevant to a bike but the general principle that, in places where wear in a joint, however unlikely, could result in sudden disconnection of the joint, some simple precaution can usually eliminate the possibility of total disconnection.

CORROSION

Corrosion of **Critical parts** is not usually much of a problem on any relatively modern motorcycle from a safety point of view, although many show cosmetic corrosion very early in their lives and are much harder to keep free from cosmetic corrosion than modern cars. From a safety point of view the problem with corrosion is that corrosion pits can form the ideal site for the initiation of fatigue cracking and the cracking may not be easily seen under the corrosion. This is a point worth considering if restoring a rusty wreck of a vintage machine, for example a set of badly rusted girder forks could be made to look as good as new, by filling the pitting with very small amounts of car body filler and a good paint job, but none of this work would help if fatigue cracking developed under the paint and eventually the forks broke on a bumpy back road.

My best (worst) story of corrosion of ***Critical parts*** *is again on a car, a Lotus Elite from the early 1970s. This was a very good car, fast, economical and much more reliable than some people believe. It had been in daily use for several years and lived outside in all weathers, so some corrosion of inaccessible parts was to be expected. The front suspension bottom link on these cars was a channel section steel member, open at the top, which carried the bottom of the spring damper unit and the bottom of the steering upright. This channel was very strong when* *in good condition but could corrode from the inside where it was not obvious to a general inspection, including several MOT tests. The bottom suspension unit mounting partially blocked the channel, leaving a small slot underneath the mounting which allowed water to drain out of the channel, but it seems that this slot became partially blocked by road debris, allowing standing water to be trapped in the channel, promoting increased corrosion. The link failed at the usual time, the middle of the night a long way from home, fortunately at low speed reversing out of a gate way. It had corroded halfway through from the inside and then failed in fatigue under the suspension unit mounting. The wheel was retained in approximately the right position by the anti-roll bar. I suppose you could classify this failure as* ***Major*** *rather than* ***Hazardous,*** *as the car continued to roll on its wheels and could be steered and braked with care, which fits the definition of a* ***Major*** *failure and given the necessary care and maintenance (i.e. regular inspection and cleaning out any road debris) it would not have happened.*

Sketch 10 shows the suspension bottom arm and suspension unit with the area affected by severe corrosion which did not come to light until after the failure of the arm.

The sketch, drawn from memory, is typical of the type of hidden corrosion site which should be avoided on ***Critical Parts****.*

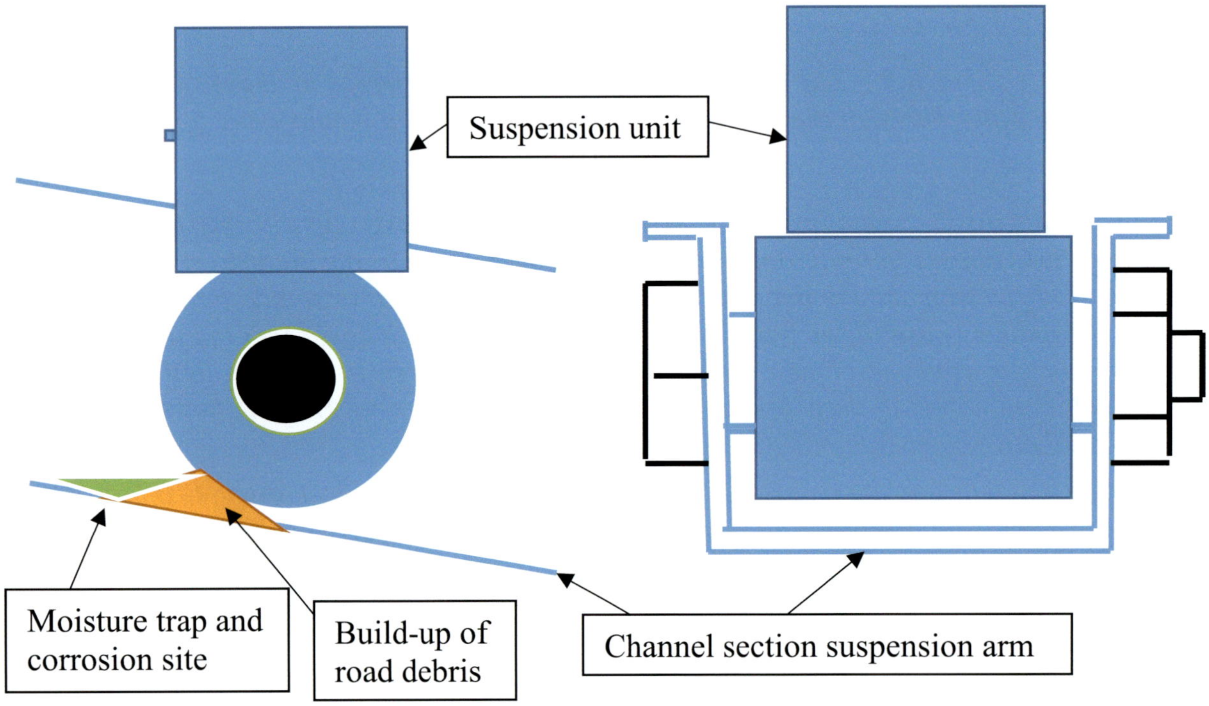

Sketch 10 – Difficult to see corrosion.

In this case, replacing the link with a new one did not seem the best solution so we made two stronger ones from square section steel tube with the same dimensions but using thicker wall tube, slightly more un-sprung weight but no doubt still going strong somewhere today.

This memory brings to mind another remembered quote from one of the previous generation of my family, who was a "tackler" (a job we would now probably call a maintenance engineer) who was close to retirement in the textile industry when I was a teenager. He was a man of few words but vast experience over many years keeping everything running often in difficult circumstances.

"If you mend broken rubbish with new rubbish, you've still got rubbish."

In other words, if you replace failed parts with identical replacement parts, without understanding the reason for the failure and doing something to improve the situation, you are likely to see the same failure repeated. As true today as it was with the steam engines, flat belt drives and looms of the old cotton mills, which were becoming a bit tired in the early 1950s after running through the war years with a minimum of care and maintenance based on "make do and mend" in the absence of a supply of new parts.

In the case of our one-off motorcycle, corrosion is not likely to be much of a problem providing that we use suitable coatings for any corrodible materials and seal both ends of all hollow parts such as tubes to avoid ingress of water. It is worth noting that magnesium is particularly prone to corrosion and, in some cases, to "intergranular" corrosion which may not look particularly severe on the outside but could have penetrated a long way into the material, forming ideal sites for onset of fatigue cracking. I would avoid magnesium for one-off **Critical parts** unless weight reduction was absolutely essential, e.g. perhaps for racing wheels, in which case the best available protective coatings should be used together with regular inspection for corrosion or cracking.

INSTABILITY

Instability is not really a subject for failure analysis, but is a design consideration. It is worth mentioning here as it is definitely safety related and could result in a **Hazardous** situation.

It seems to me that instability could result from one of two causes: instability as a result of forces generated at the tyre to road contact point, which you could call mechanical instability, or instability resulting from aerodynamic effects.

"Mechanical instability" is a low frequency vibration of the whole bike and, in former days, was known as a speed wobble. As I see it, this results from the balance of forces applying torque to turn the steering. With no manual input from the handlebar, this torque comes from three separate effects:

- The inclination of the steering head results in a tendency for the bike to fall onto full lock if disturbed because the height of the bike's centre of gravity is lower on full lock. Increasing the steering head angle from the vertical increases this effect.
- When moving, the above tendency is counterbalanced by the effect of "trail", where the point of tyre contact with the road is behind the intersection of the centre line through the steering head bearings with the road. Thus, any turning of the steering when moving produces an instantaneous side force on the tyre which pushes the steering back towards the straight ahead position and simultaneously produces a counter-steer effect tending to lean the bike in the opposite direction.
- The situation is further complicated by the gyroscopic effects of wheel rotation which basically resist any turning of the steering or changes in lean angle. The effect increases rapidly with increasing speed and, at very high speeds, could make initiating a turn very difficult. Very light wheels are the only answer to reducing gyroscopic effects, if they become a problem, although this is unlikely except at racing speeds.

In normal running, sufficient trail and the increasing gyroscopic effects will keep the bike stable and upright but if sufficiently disturbed, for instance if the front end lifts slightly, perhaps due to road bumps or hard acceleration, and lands with the wheel slightly out of line with the direction of travel, the effect of trail will be to make the steering snap back into line. Simultaneously, the momentary counter-steer effect will cause the beginning of a lean and consequent turn in the opposite direction. If this snap back effect is violent enough, the movement could over-swing past the straight ahead position, reversing the effect of trail and counter-steer and repeating the cycle backwards and forwards. On a well-designed bike, this effect should die out after a few cycles, helped by the steering head angle which effectively lifts and lowers the bike centre of gravity as the steering is turned, and perhaps by a steering damper, if fitted, which is intended to provide resistance to slow any rapid movement of the steering.

A few years ago, there was some interest in adding weights to the ends of the handlebars to increase resistance to sudden unintended steering angle changes and reduce the natural frequency of oscillation of the steering by increasing the rotating inertia around the steering axis.

I believe that the potential benefits from this are quite small, and it could be detrimental to recovery of the situation if a severe oscillation did become established. A hydraulic steering damper, which reacts to the speed of movement of the oscillating steering rather than just the initial acceleration from the centre position will always be a more effective solution to a "lively front end" than bar-end weights.

If the design of the steering system is not the best, the side-to-side oscillation of the steering could build up, causing an increasing weave known as a tank slapper, which could be impossible to control; a definite **Hazard** (think of a supermarket trolley wheel which generally has a lot of trail but a free running vertical steering head on each wheel. With nothing to damp the oscillation, they can very easily develop a severe wheel wobble).

The solution based on the vast majority of stable machines, seems to be that the steering head should be angled between about 25°± 5° to the vertical and trail should be around 3.5 to 4.5 inches on a medium weight bike of reasonable conventional design.

There are many other factors which could have some bearing on the potential for instability, such as weight distribution, tyre sizes, profiles, and pressures, so the best approach for a modification or unique design is to start from a point similar to a comparable production machine. Then if any sign of instability is detected, don't ignore it, but experiment with various changes to try to eliminate it. Steering head angle can be varied by raising or lowering the back end, with 1 inch roughly equivalent to 1° of angle change on a typical design. The only way to vary trail significantly with a conventional telescopic fork is to make or obtain new fork yokes which move the fork legs, forwards or backwards, relative to the steering head. Other suspension systems such as leading links can be designed to allow easy adjustment of trail. Fitting a steering damper to cover up the problem is, to my mind, a last resort on a 2 wheeled bike *(a trike with 2 wheels at the front is different in this respect and a steering damper is likely to be essential as we will get to in section 5).*

In my experience, a steeper, nearer vertical, steering head angle will produce a quicker steering machine which may not be very stable, perhaps because turning the steering results in less change to the height of the centre of gravity and thus less resistance to rapid movement. The opposite will result in a very stable machine which may be a bit reluctant to turn sharply at high speed. At very low speed, with minimal gyroscopic effects, a very inclined steering head tends to make the bike seem to "fall into a corner" perhaps more rapidly that you are expecting, requiring a quick correction or a foot down to restore the intended line.

The influence of trail on stability would seem to be less clear cut. If the front wheel became out of line with the direction of travel for any reason, a large trail dimension would result in a larger initial torque impulse to the steering, all other things being equal, because of the longer movement arm between the tyre contact point and the steering axis, but the increased self-centring effect could be beneficial in restoring stability providing that the effect is not so violent that it introduces an over swing as mentioned above.

Somewhere in the middle of the normal range would be the best starting point for any unique machine. On all the machines where I have been able to select the steering head angle and trail (apart from the supercharged Velocette, Photo 13), I have settled for 26° head angle and 4 inches of trail with the bike static and unloaded. I offer no guarantee, but none of them have had any problems with this type of instability and I have never had to resort to a steering damper.

I have only experienced something approaching a full-blown tank slapper once, many years ago on the Vincent on a straight back road with many potholes and bumps in the surface. I was perhaps going a bit too fast for the conditions being overconfident of the Vincent's stability. This was before its conversion to the Norton based front end (Photo 7 shows a similar unmodified Vincent). I stayed on-board and, more by luck than management, the oscillation died out after a few cycles. I never really discovered the cause at the time, as the Vincent, in standard form, was usually a very stable machine with something like a 30° steering head angle and more trail than most large capacity machines. It never happened again. Thinking about it today, you could speculate that the instability may have been initiated by running over a sharp-edged hole in the road surface with the front tyre hitting the edge of the hole at an angle pushing the steering out of line with the direction of travel. The long trail could then have caused a violent return overswinging the centre position and reversing the effect.

Photo 7 – The Paul Adams Vincent Black Shadow.
(Photo reproduced courtesy of Paul Adams)

Photo 7 above shows a magnificent Vincent Black Shadow owned by Paul Adams of the Vincent owners club. This bike is very similar to my Rapide before it was modified, although mine was never as tidy and beautifully finished as this one.

The second possible cause of instability is the effects of the air pressures exerted on the bike when travelling at high speed. This obviously increases with increasing speed and is unlikely to cause too much of a problem at most normal road speeds.

Aerodynamic instability is a big subject, beyond the scope of this book, but basically, there is a point called the centre of pressure where the sum of all the air pressures exerted on the bike, as a result of forward speed and any side winds, can be assumed to act. This point tends to move forward as speed increases but its position relative to the centre of gravity controls the aerodynamic stability or otherwise of the bike.

Aerodynamic forces on the bike can be divided into drag, lift, and side force.

Drag, sometimes known as air resistance, is the rearward force acting to slow the bike and combating it uses most of the engine power needed to maintain a speed. For this reason, any change which reduces drag, such as adding a well-shaped fairing, is a good thing from the point of view of performance although it may have disadvantages in other respects. Another feature of drag is that it causes weight transfer from the front to the rear wheel because its line of action through the centre of pressure is above the tyre contact points with the road. The amount of weight transfer obviously depends on speed and should not be too significant at normal road speeds. We can therefore conclude that drag is largely a performance issue and is unlikely to have any potentially **Hazardous** effects on our modified road-going machine.

To put some approximate numbers to drag forces, the total drag on a naked machine plus rider would be expected to be around 100 to 150 lbs at 100 mph, depending on total frontal area and the overall size and shape of the machine and rider. This drag would absorb between around 25 and 45 horsepower to which the power absorbed by the mechanical operation of the transmission and rolling resistance of the tyres must be added. Drag increases approximately with the square of the speed, and the corresponding power requirement with the cube of the speed, so at higher speeds the power requirement rises very rapidly. Weight transfer to the rear, resulting from drag, also increases rapidly and could be partially counteracted by aerodynamic front downforce, (negative lift) generated by a well-shaped front fairing or by a long rear suspension swinging arm, effectively moving the centre of gravity forward. *(Nothing comes for free, and a long swinging arm has advantages in stability at very high speed and in keeping the front down during acceleration but tends to make the bike front heavy in low speed manoeuvring and make losing the front more likely on wet low speed slippery corners.)*

Lift is the vertical force generated by the airflow over the bike and can be positive, upwards, or negative, downwards. An aircraft uses positive lift, mainly generated by the shape of the wings but, for any road vehicle, negative lift is desirable as it helps keep the wheels on the ground. This is particularly true at the front of a motorcycle as negative lift can go some way to combat the weight transfer due to drag. At normal road speeds, lift should not be too much of a problem for our modified motorcycle, as it is not likely to be large compared with the weight of the machine and is unlikely to have any **Hazardous** effects. However, any change which causes or increases positive lift should be avoided, particularly at the front of the bike. I would therefore suggest that if the modification involves changes to the front bodywork or fairing, some thought should be given to the effect on lift, particularly if high speeds are intended.

Side force is the effect of air pressure on the sides of the bike and will be zero when travelling straight on an absolutely wind free day. Any side wind blowing across the path of the bike will cause a side force tending to lean and steer the bike away from the wind direction and some movement of body weight or steering correction will be needed to maintain a straight course. The turning effect of side force, applied at the centre of pressure, will increase if the centre of pressure is moved forward. The main problem comes if the side wind is not steady but comes in gusts which require constant adjustments to maintain a straight course and could get to be difficult to control in the worst case, possible becoming **Hazardous** if a sudden gust causes an unintended turn at the wrong moment. Therefore, any change which may move the centre of pressure forward, or increase side area, such as more front bodywork or a dustbin type fairing, requires some thought as to its effect on aerodynamic stability. In general, flat "slab sided" panels should be avoided as much as possible, particularly at the front end, in favour of more rounded shapes in cross section.

As an extreme example to illustrate the effect of the position of the centre of side force pressure, the vertical tail fin on an aircraft is mainly intended to move the centre of side force pressure rearwards to improve directional stability. If the aircraft is disturbed in flight such that the aircraft centre line gets out of line with the passing airflow, air pressure on the tail fin will rotate the aircraft back into line with the airflow. If the fin were at the front, moving the centre of pressure forward, this same air pressure on it would push the aircraft further out of line making the situation worse. Sketch 11 illustrates the effect.

A motorcycle is in a somewhat better position than the aircraft as it also has the wheels in contact with the ground to help maintain straight line motion, but the same principle applies. That is that moving the centre of pressure forward will make it more difficult to maintain a straight course in gusty conditions and in the worst case could result in an increasing weave which may be impossible to control.

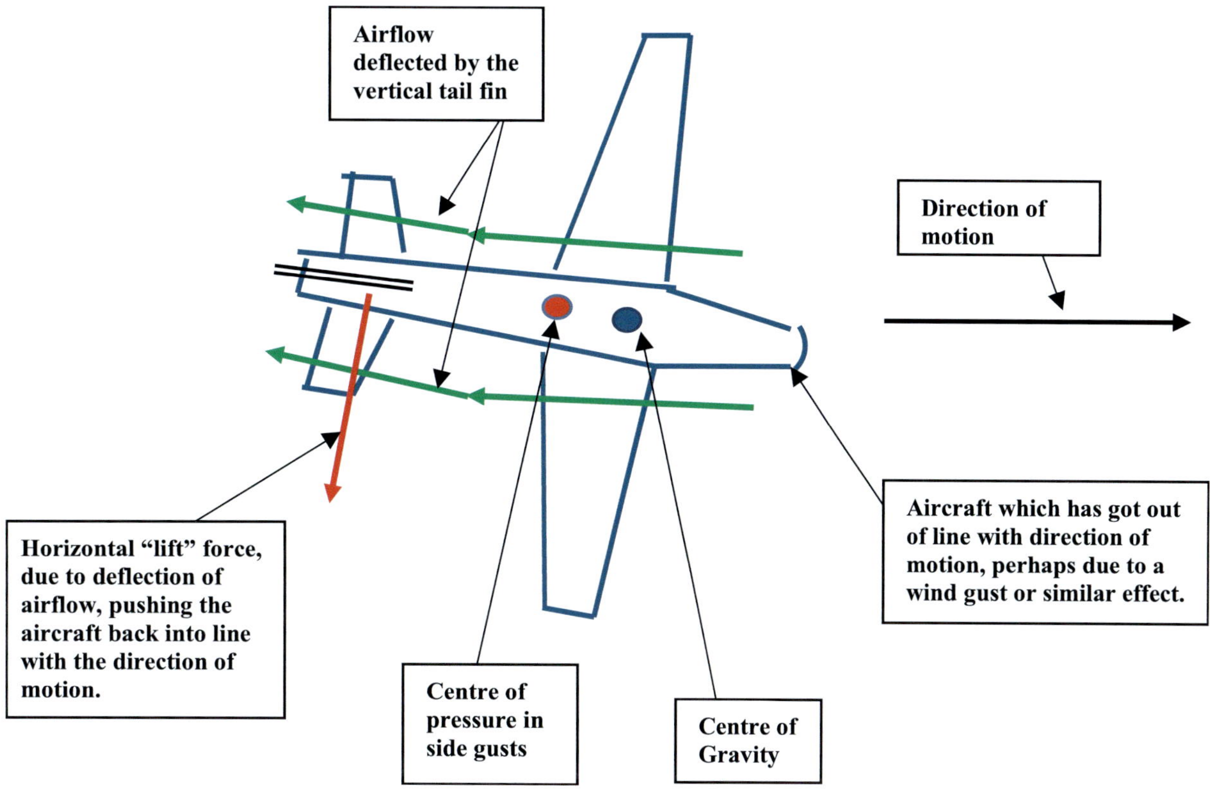

Sketch 11 – Illustrating the effect of centre of pressure on stability.

Chapter 7 – In Conclusion

The methods described in this book are all tried and tested in the aircraft industry and the safety record of current aircraft shows that they are effective *(although recent events have shown that even a major aircraft manufacturer can occasionally get it very wrong with Catastrophic results).*

They really amount to logical application of common sense, rigorously applied to every aspect of the design, ensuring that no potential **Hazard** is missed.

We can achieve a similar result with our one-off motorcycle, without too much difficulty, if the principles of failure analysis are born in mind when deciding how to approach each change from standard. This assumes that all standard manufacturers' parts are satisfactory from a failure analysis point of view as shown by good experience in normal use with a large number of riders. There is, however, no harm in casting an eye over the standard parts around the area to be modified to ensure that there are no places where the consequences of a failure can be made less **Hazardous** by some small change, even if the probability that type of a failure happening is low.

After all the above is covered, there is nothing like a bit of good old fashioned engineering judgement, based on experience and knowledge of what has been successful before, as a final check. Another old engineer's saying covers this point exactly:

"If it looks wrong, it usually is wrong and we have probably missed something, but if it looks right, there is good chance, but not a certainty, that it is right."

Finally, some readers may think that I am overly pessimistic and old fashioned in constantly looking for the ways things can fail, rather than believing in my ability to always get things right first time, but they should consider that even the smallest manufacturer does hundreds or probably thousands of miles of testing and introduces modifications to fix any problems found before releasing their product for sale and general use. We special builders can usually only do a small fraction of that sort of work so are much more reliant on getting it right first time.

(Perhaps an overconfident Triumph owning reader who criticised the first edition as too negative, calling it "Elf and Safety Rubbish" somehow missed that point? I hope he does not find it the hard way.)

Part 3 of this book looks at the different sections of a motorcycle and gives a general idea of what to look for. All designs are different in some details so the list is not exhaustive but hopefully will encourage the practice of taking a critical look at each non-standard part of your creation just in case you have missed some point which could result in a bad end to the project.

Before we get to part 3, I have included part 2 with some more photographs of a few of my various efforts over the years, not forgetting 4 wheels as well as 2 and 3. All were produced without spending vast sums of money and, in the early days, with very little equipment, although I have collected much more over the years. It just needs a bit of imagination and a lot of time and effort.

Perhaps they may inspire a few people to have a go at making something different.

PART 2

A Selection of Other Specials Built Over the Years

Photo 8 – The Excelsior Manxman.

This machine was my first attempt at building a special in 1956/57 and was based on a 1936 350 cc Manxman which had seen better days when I bought it for £12, if I remember correctly.

The concept was to make the nearest thing to a Manx Norton that I could afford and overall, it was counted a success although it did, at one point, suffer the potentially ***Hazardous*** *failure related to the rear brake anchor referred to in Part 1, Chapter 1 – Principles.*

The forks were from a Matchless and were available at low cost at the time, surplus stock from the large number of surplus army despatch riders' bikes. The rear suspension was homemade with a swing arm and rear sub-frame welded up from steam pipe and bolted to the Excelsior front frame.

It did have a silencer added for the road but had no lights or any other electrical system other than the magneto ignition. (The supporting brick was eventually replaced with a prop stand.)

Photo 9 – The first Triton.

This was my second attempt after the Excelsior, aimed at better performance and handling than the Manxman and some ¼ mile sprints in addition to use on the roads.

It was based on a new Norton featherbed frame which came from a bike where the engine and gearbox etc. had been taken for use in a 500 cc formula racing car — a common practice at the time as the Manx engine was the best engine available for these cars. The engine was always believed to be a Triumph GP 500 cc twin which had forward facing exhaust ports as distinct from the angled ports of the road-going twins. It could have been, although very few GPs were produced and there was an aircraft auxiliary engine of similar design used to power generators in some WW2 aircraft and available on the surplus market at the time. The gearbox was Triumph and the rest of the equipment a mixture of Norton and Triumph parts so not much original thought required.

However, it was very light compared with production machines of the time and ran very well with exceptional handling. It certainly kept me amused for a while and was something different before the cult of Triton building became common and almost productionised as it continues to this day.

Photo 10 – The first Triton (2.)

I have included this photograph, which has not been cropped or manipulated in any way, as it was considered at the time to be 1000 to one chance of getting a full frame picture with the bike passing, at close to its maximum speed, during a bit of private testing on the public roads in preparation of a ¼ mile sprint.

Good days before universal speed limits, speed cameras and endless traffic, (or expensive riding gear).

It was taken by taken by a friend, who was a good amateur photographer standing at the side of the road with a single shot film camera.

Photo 11 – The Vincent Rapide.

This bike was a 1953 1000 cc Series C Rapide and was my everyday transport in the early 1960s. It was modified over a long period by gradual improvement mainly to improve "rideability" in general use. The forks and front wheel were Norton Dominator, and the rear suspension modification replaced the Vincent single suspension unit with a conventional twin shock system, providing better dampers but retaining the stiff Vincent triangulated swinging arm. The rear sub-frame was homemade and was arranged to raise the back end to steepen the steering head angle, reducing trail and making the bike much more agile with a more modern feel to the handling. The sub-frame also supported the seat and rear mudguard directly in place of the Vincent system of support from the swinging arm for the rear of the seat and the mudguard. The Vincent servo clutch was replaced with a Norton clutch for more consistent operation. The modified bike was, in terms of miles covered and years of regular use, my most successful project. I used it for over 10 years with few problems. Fellow Vincent owners were sceptical about the modifications at the time but generally approved after they had ridden it. It was proved to be faster than a new Triumph Tiger 110, one of the best sports bikes of the day, on equal terms on an (unofficial) standing start hill climb test but was not as fast as the best Vincents around at the time which were fitted with all the available engine modifications.

It did everything from long distance daily commuting and holidays to ¼ mile sprints and the 1960s equivalent of track days and, with the one exception referred to in part 1 chapter 6, it always got home in one piece under its own power.

Photo 12 – The second Triton.

This one was not all my own but was obtained, part finished in part exchange for some parts, so I had little influence on the main features of the design. It had a standard Triumph Tiger T100 engine with a few tuning modifications and a Norton gearbox with the engine inclined forward in a Norton Dominator type featherbed frame. The oil tank was in the rear section of the fibreglass petrol tank which I suppose saved a bit of weight, but not enough to compensate for the much heavier frame compared with the first Triton. The oil was separated from the petrol by a double wall with an open space between the walls so there was no chance of one leaking into the other, but the outside of the tank was always hot and slightly oily. Not a good idea.

The bike had marginally more power than the first Triton but was heavier so was not much different in performance. The inclined engine made no measurable difference to the handling or anything else. I did not keep it long after finishing it off as I still had the Vincent which was a much more versatile machine.

Photo 13 – The supercharged Velocette Valiant.

Built in the early 1970s as an experiment.

The concept for this bike was to make a very light machine with performance similar to the 250 cc two strokes and 350 cc four strokes of the day using a small 200 cc four stroke twin with a centrifugal supercharger. It was also my first attempt at doing something different with the front suspension system when the vast majority of production machines had settled for telescopic forks.

It never achieved the desired performance, due mainly to the limitations of the supercharger Vee belt drive although it did have a good performance boost at high rpm. The leading link front suspension was a success and the anti-dive geometry coupled with the low overall weight and the large diameter Honda front brake gave by far the best overall braking performance with drum brake I had achieved up to that time. The 4 leading shoe double sided brake fitted to the Domiracer replica, Photo 20, at a much later date was more powerful but tended to be a bit harsh at low speed.

It was perhaps a concept in advance of its time with the current revival of interest in small capacity supercharged engines, particularly in the car world.

Photo 14 – The horizontal Triumph (1).

This machine, shown above with the tank, seat and side panels removed, was built to retain all the good features of the Velocette (Photo 13), while incorporating lessons learned on that machine, and improving the performance by the simple means of using a larger engine.

The front end was almost identical to that used on the Velocette with the leading link forks and the same type of drum brake and anti-dive geometry and with a homemade steel spine frame and aluminium plate rear sub-frame.

The engine and gearbox were Triumph Tiger 100, chosen as the lightest 500 twin available, with the oil system modified for horizontal running. The primary chain case was from a BSA. The complete engine/gearbox assembly was mounted from a single pivot point at the rear and suspended from a rubber mounting above the crank case. The cylinder head was directly air cooled and an air scoop was fitted to direct cooling air onto the cylinder barrel. The oil tank was in the frame and the air scoop below the engine incorporated an oil drains tank to drain the exhaust rocker boxes. This oil then flowed into the bottom of the crankcase for return to the tank.

The supercharger was a roots type driven by a toothed belt. It was mounted from a pivot point on the frame and connected to the engine by an adjustable length link to ensure constant belt tension when the engine moved on its flexible mounting. A supercharger bypass valve was fitted, open below 75% throttle opening, to reduce the power lost as a result of driving the supercharger when it was not required.

The double swinging arm rear suspension was used to maintain a relatively short wheelbase, compensating for the increased length of the engine/gearbox assembly. It also resulted in almost constant chain tension as the suspension moved, which was an advantage of the arrangement.

Photo 15 – The horizontal Triumph (2).

The bike performed well from the start with the light weight, low centre of gravity and suspension, proved on the Velocette, giving good handling and braking. The engine also was fairly trouble free apart from a few necessary changes to the oil system, mainly to improve oil feed to the supercharger gears and drainage from the rocker boxes. Achieving a satisfactory fuel/air mixture at all conditions with the single SU constant vacuum carburettor was also a challenge in the days before rolling roads and exhaust gas analysers were easily available outside the industry. The first attempt, two into one, high level exhaust worked well but completely spoiled the look of the bike so was changed to two separate exhaust pipes and silencers.

The fuel tank, seat, exhaust, and side panel went through a considerable number of alterative versions to improve the visual appearance before I settled for the final version shown.

(Shown with my wife Jennifer who is not too keen on bikes.)

This is the only one of my specials which was covered in the motorcycling press. A picture of an early version appeared in Motorcycle News (8 July 1981) and a full description of the final version in Classic Bike (January 1985).

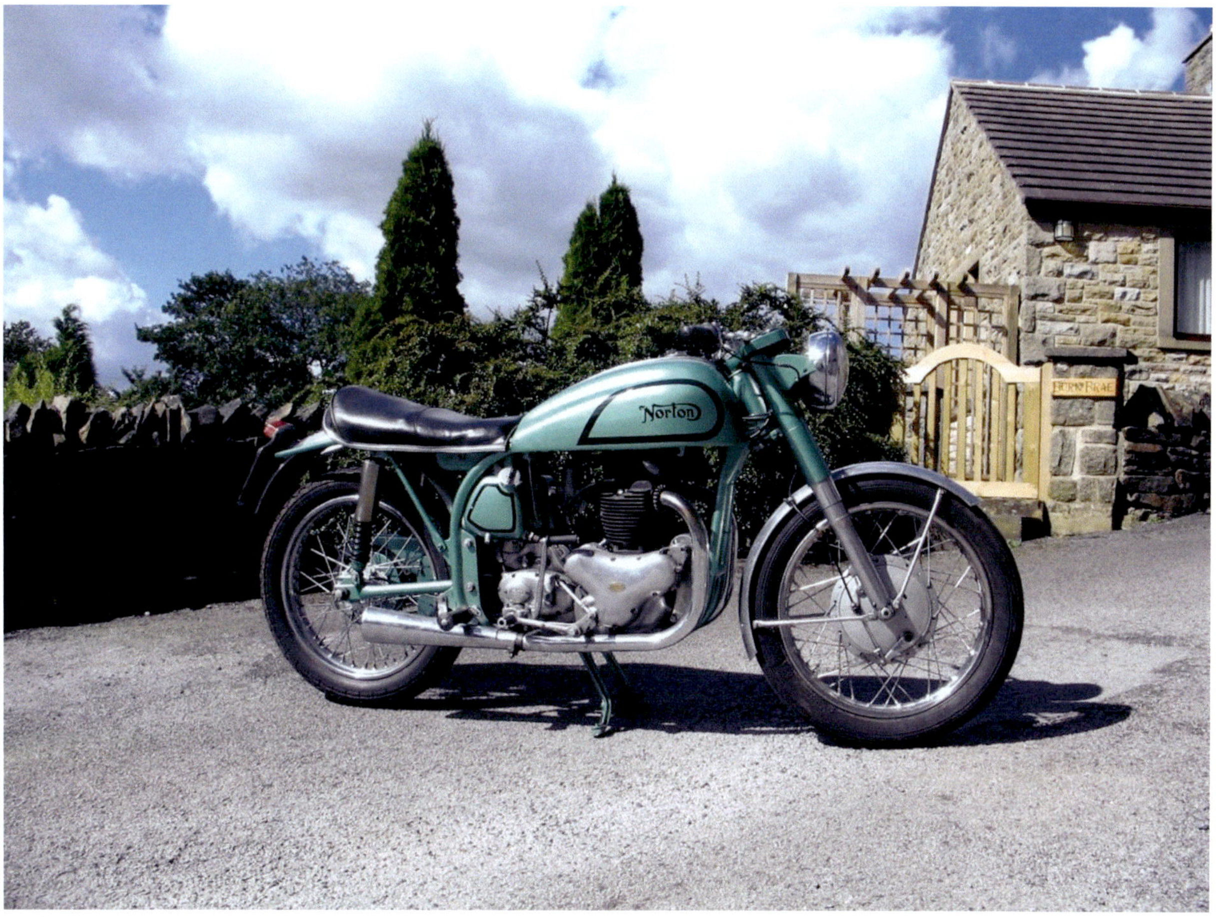

Photo 16 – The 1954 Norton Dominator 88.

This bike is one of the first featherbed Dominators and came to me in very run-down condition in the 1970s. All the main parts were original and there seemed to have been no major work done since new. These days someone would restore it as original and sell it for a vast sum, but I decided to rebuild it changing as little visually as possible but updating it, reducing weight, and improving it wherever possible. The frame is much lighter than the standard road-going featherbed and has a flat on the inside of the right-hand top tube, which is intended to clear the camshaft drive on International and Manx Nortons, but it had been fitted from new, perhaps indicating some Clubman racing history.

The main changes are, from front to back, are as follows.

Later full width front brake, stainless rims and spokes, the stronger right hand fork sliders fitted to both sides, lightweight mud guards, BSA flat back chrome headlamp, Vincent type straight stainless handlebars, aluminium engine plates, 12 volt dynamo conversion, later Dominator aluminium cylinder head, BTH electronic magneto, high flow oil pump, belt primary drive and diaphragm spring Commando type clutch, rear set footrests with 1 down 3 up gear operation, new battery box for Burlin 12v long life battery, folding kick starter, Matchless silencers, green paint scheme because I like green bikes, (and perhaps a few other changes which I have forgotten).

No radical changes and none which increase the probability of a ***Hazardous*** *failure, but the overall result is a very nice bike to ride, not fast but smooth and reliable with exceptional handling and it has had very few problems over the years. One to keep. I still have it after 50 years and use it occasionally on nice days in summer.*

Photo 17 – The Feet Forward.

Built in the 1990s.

The concept for this project was to experiment with the idea of a low seat feel forward machine which would reduce aerodynamic drag because of reduced frontal area and also provide more comfort over long distances with reduced wind pressure on the rider and a solid back rest.

The bike was based on a Yamaha RD400 which I had used in standard form for some time. Almost all of the RD400 was used unchanged with an extra box section extension moving the swinging arm pivot point backwards to provide room for the low, car type, seat. A spring-loaded tensioner was added to the bottom run of the new long final drive chain. The handlebars were at the back of the petrol tank, linked to stub bars on the fork crown by two ball jointed push/pull rods. The footrests were forward mounted with the rear brake and gear lever connected by rods to the unchanged brake master cylinder and gearbox. The lever seen beside the cylinders operated the kick starter by hand, easy with a two-stroke twin but a larger engine would have needed an electric starter.

The two go-kart wheels under the seat were fitted, after much experiment, to ensure safety in the event of grounding of the forward footrests at high lean angles for reasons explained in the book. This was a precaution which was never seriously called into action during my riding but was good failure analysis practice. The notes below Photo 18 cover the riding experience.

Photo 18 – The Feet Forward showing the riding position.

The whole objective of this bike was to see if feet forwards, then widely discussed in the motorcycle press, really were the future. I think subsequent events have shown that the answer is NO. They are very good for long, high speed journeys on open roads, with good comfort and much less power required to maintain a speed, and they corner well in these conditions feeling totally secure at high lean angles and very stable in all conditions. The downside is that they are less good in confined situations at low speed where much road riding is done these days. The main problem is the rider's lack of confidence that he can keep the bike upright during low-speed manoeuvring because the riding position puts his legs at an angle which makes it almost impossible to save any unintended high lean angle by putting a foot down. The other problem is that inability to move the rider's body weight about makes the likelihood of an unintended lean angle much greater. The go-kart wheels help a bit on the confidence front but not enough to make it feel comfortable.

I am unlikely to do another one but if I did, one idea which struck me was that it would be possible to make a system where the centre section of the car type seat could be raised up to something like a normal motorcycle riding position, hydraulically or electrically, at the touch of a button for confined spaces and lowered for faster running. The simultaneous re-positioning of the footrest and controls to suit would require a bit of thought but nothing is impossible. That could give you the best of both worlds.

Photo 19 – The Nor-Yam.

This bike was made at the time when it seemed that two strokes were the future of motorcycling and combined one of the best two stroke engines of the time, the Yamaha RD400E, with a Norton featherbed frame and RD400 forks, wheels, and brakes. It was a very successful combination and was the lightest and fastest of all my Norton framed bikes.

The engine incorporated the Yamaha rubber bushes in the engine mountings which necessitated the extra vertical support tube shown in front of the cylinders to prevent the engine creeping downwards at the front and overloading the bushes.

The biggest problem was the "Allspeed" exhaust system, essential for performance but impossible to fit in the intended position. It was eventually persuaded to fit in the position shown by exchanging the left- and right-hand pipes and adding a few judicious dents to clear various obstructions.

Photo 20 – The Domi-Racer.

This bike was inspired by the 1960s Norton works racers which were intended to take over from the single cylinder Manx but were never fully developed by the company at the time.

The bike is based on a much modified 500 cc Dominator 88 engine in a wide-line featherbed frame with a toothed belt primary drive and diaphragm spring type clutch. It has no electrical system apart from a self-generating electronic ignition unit mounted on the end of the crankshaft. The belt cover is homemade from aluminium sheet and most of the remainder of the bike is assembled from proprietary parts from Norton specialists. The front brake is a double sided 4 leading shoe drum brake of unknown origin, but it is very effective, if a bit too heavy.

Altogether, a good "Sunday afternoon special" and remarkably similar in principle to the first Triton, although separated by about 45 years and made from much better quality parts. It shows that my taste in bikes, based on lightweight, simplicity and enough but not excessive power, has not really changed over time.

Photo 21 – The electric monkey bike.

This bike was built for my Grandson at about age 7, to introduce him to powered two wheelers. It was based on a standard 50 cc Honda bought in run down condition and rebuilt unchanged except for replacement of the engine and gearbox with a 24-volt ½ horsepower electric motor with a single speed, 2 stage, chain drive using bicycle chain. The bike was geared to have a top speed of about 10 mph which resulted in plenty of low-speed torque. Electrical power came from 2 standard 12-volt motorcycle batteries, one in front of the motor and one alongside. Power was controlled by means of a twist grip operating a proprietary electronic control system which gave good progressive control of torque to the back wheel.

Overall, it was counted a success providing him with entertainment while learning the basics of balance, cornering, and riding over obstacles. It had a hard life riding up and down steps, over jumps and kerbs and gave very little trouble apart from the occasional blown main fuse. I also learned a little bit about the practicalities of electric bike design which may come in useful in the future.

Photo 22 – The Kawasaki ZXR 400 L9.

This bike was used for 17 years from new in 2000, at all times when something more unusual was not going. I resisted the temptation to do major modifications to it as, from the handling and braking point of view, it would be hard to improve on the standard bike and it has more than enough power for any reasonably sane road use. I have just civilised it a bit, more suited to a rider long past his best, by raising the handlebars grip positions by 4 inches and lowering the footrests by 1 inch to give a more comfortable riding position and fitted a softer rear spring more suitable for solo road use. I have also repositioned the seat which as standard slopes down toward the front causing the rider to slide forwards into the tank, most uncomfortable.

The reason for including it in this book is to show the crash bobbins fitted after an unfortunate experience ended with the bike sliding up the road on its left-hand side. This quickly rubbed through the thin aluminium alternator cover and deposited all the engine oil on the road. A real ***Hazard*** *to other road users.*

When repaired, I fitted 3 crash bobbins to each side with one positioned, as shown by the arrow, to prevent the generator cover from being a first point of contact if the bike is on its side. Hopefully they will never be needed but good failure analysis rules say that if a ***Hazardous*** *failure is identified, however unlikely, action must be taken to prevent it occurring. Another lesson learned for any future two-wheeler.*

However, they will never now be needed as the bike has retired from two wheeled use and is now a trike as described in part 5.

Photo 23 – The modernised Dominator.

This bike is still in the early stages of development and is an attempt to introduce some more modern features into a 1960s Norton Dominator without altering the classic bike riding characteristics too much.

It was largely built from parts left over from previous projects so I did not destroy an original classic, which should please some Norton purists.

It is based on a wide-line featherbed frame increased in width at the lower rear corner to accommodate an increase width modified Norton swinging arm. The wheels, forks and brakes are Kawasaki ZXR400 and the 6-speed gearbox is from a Yamaha RD400. The engine is a much modified 500 cc Norton Dominator with chain drive to a cross shaft running in the original Yamaha crankshaft bearings to drive the RD400 primary gears and clutch. A homemade, machined from solid, aluminium primary chain case encloses the chain and also provides a mounting for an electric starter. The starter drives the end of the cross shaft through 3 gears and a one-way clutch.

The bike also has electronic ignition, indicators, LED lights and electronic instruments powered by an Electrex World combined alternator and ignition unit which has proved to be a very good option in the somewhat crowded world of electronic ignition systems. A homemade two into one exhaust system using smaller diameter stainless tube and a lightweight silencer from an off road machine saves a lot of weight and is no noisier than the original Norton system. There are also Matchless, BSA, Ariel, and Vincent parts in there somewhere in addition to many other homemade parts.

Since completion this bike has only done a relatively small mileage as a result of lockdown restrictions and other projects but has been trouble free to date. The only change made was to go to smaller section, higher profile tyres as the low profile ZXR 400 tyres made the bike much less agile than the standard Dominator. This change seems to have fixed that problem.

Photo 24 – The Mk 1 reverse trike.

The concept of this bike was to experiment with the somewhat neglected field of 3 wheelers while keeping it more motorcycle than car, and to produce something which was easy to push around for someone past their better days from the point of view of strength. (It cannot fall over in the garage.) It is based on a 400 cc Honda Bros which I had used a few years ago but had not ridden for several years. If you are experimenting with a new chassis/frame/suspension design, it is a good plan to use a reliable standard engine and transmission. Changing everything at the same time is a recipe for problems and delays.

To simplify the construction of this first attempt, the front sub-frame and front wheel arrangements were based on a modified car sub-frame reduced in width to suit. More details of the advantages and disadvantages of this approach are discussed in Part 5.

I chose to have two wheels at the front because I always instinctively felt that a conventional trike was the wrong way to go. The centre of gravity is generally too far back for aerodynamic stability at speed and there is often insufficient weight on the front wheel to ensure that the machine will not understeer badly, particularly in wet or slippery conditions. Another point, with the failure analysis in mind, is that, with two wheels at the back, the back suspension and wheels are ideally placed to run over the rider or passenger's legs if they happen to put a foot down at the wrong moment and it catches on some road irregularity. Could be nasty.

After the initial road testing and some adjustments, the machine has performed well with few problems and is generally reliable. It has now been on the road for over six years including several visits to the IOM and can be counted as a success. Its only real fault is too much roll during cornering. With the Mk 2.5 now running, an upgrade to high roll centre front suspension is on the list of jobs to be done.

Photo 25 – The Mk 2.5 reverse trike.

This machine is my second attempt at a 3-wheeled conversion, the Mk2 and Mk2.5 versions are based on the Kawasaki ZXR 400, Photo 22. I have tried to incorporate all the lessons learned from the first attempt, Photo 24. In place of the modified car front sub-frame, they include an individually made aluminium structure with a suspension designed to give a high roll centre to minimise roll when cornering as shown above. It also uses the lightweight Kawasaki front brakes.

The bodywork is mainly aluminium with some fibre-glass parts and is more minimalist than the first effort, mainly for appearance. The final Mk 2.5 version includes two stub wings in front of the front wheels, again largely for appearance, but the small front downforce generated at road speeds does no harm.

It is now on the road and has proved to be considerably better than the Mk 1. On normal relatively smooth roads, it is very stable with consistent handling and almost zero cornering roll with no change from the initial design. On very rough surfaces, finding a good compromise for front suspension spring rate has proved difficult, with the lower sprung and unsprung weights, but a reasonable compromise has now been found. The only reliability problems have been with the unchanged 20-year-old bike parts which, to date, have not been as reliable as the Bros based Mk .1. Time passes when you are stuck on one leg, so hopefully it is now fully sorted for future riding seasons.

Part 5, included in this 2nd edition, attempts to cover the development of these 3-wheeled machines in more detail as guidance for anyone contemplating a similar project. It is well worth the effort as the riding experience is different but equally satisfying when compared with two wheels.

To prove that I have not entirely neglected the 4 wheeled world I have included two of my 4 wheeled efforts. Whether on 2 or 4 wheels, the same principles of failure analysis apply. The only differences are that the types of failure and potential Hazards are different.

Photo 26.1 – The twin engined Hardaker Mk 1.

Photo 26.2 – The Hardaker Mk 2.

The Hardaker Mk 1: The Twin Engined Car

This was the most radical of my 4 wheeled efforts, built in the late 1960s and early seventies. It had two 850 cc Mini engines, one in the front and one in the back to provide 4-wheel drive. I had long been convinced that 4-wheel drive was the way forward for sports cars, as there was a limit to what could be achieved by just adding more power to a 2-wheel drive car, particularly in the days when tyre grip was not as good as today. Twin engines seemed to be the easy way to achieve 4-wheel drive.

The chassis was based on a central aluminium alloy tub with a double skinned floor and two deep side sills and with a steel tube bulkhead at each end. The tub material was a high strength 16-gauge (0.064 inches) aluminium alloy sheet riveted and bonded together with almost 1000 alloy rivets and filled with rigid polyurethane foam. Two Mini front sub-frames were attached to the front and rear bulkheads complete with suspension, steering (locked at the back) and engine mounts. The larger 13-inch wheels resulted in suitable gear ratios, with double the Mini power and similar weight, with no change to the final drive gears. Drum brakes were fitted initially but the fronts were later changed to discs.

The fibreglass and aluminium upper body was made from a "Rochdale" body salvaged from a scrap Ford based special and cut into 6 sections, increases in width with a 4 inch strip down the centre and arranged so that the front and rear sections could be hinged up for access to the engines and suspension, etc.

The single clutch pedal and gear lever operated both engines together with removable pins which allowed single engine running with the other engine remaining in neutral. The throttle pedal was split into two vertically so that angling the foot allowed the ratio of power to the front and rear to be varied.

After a bit of development and a few modifications, notably to improve the linked gear change and move the fuel tank, it was good to drive, handled well and created a lot of interest. It was at its best on slippery corners where power could be applied at the rear at the apex and the front brought in as the corner opened up. (It is much more satisfying, when you get it right, controlling the torque distribution between wheels yourself rather than just pressing the throttle pedal letting a computer sort it out.)

The fuel tank was initially in the left-hand side sill, separated from the passenger compartment by a fire wall, but still vulnerable to a side impact. In line with good failure analysis practice, it had to be moved, but space was very limited. Two small tanks behind the seats were eventually fitted with a fire wall in front of them. Inconvenient for filling and using up the last bit of available space but a Hazard eliminated.

Its problem as a road car was total lack of carrying capacity for anything other than the 2 people so in the end it had to go. All the upper bodywork and the front engine were removed, and the car gradually evolved through several versions during the 1970s and early 80s into the Hardaker Mk 2, ***Photo 26.2.***

Perhaps the twin engines driving both pairs of wheels was another idea a bit in advance of its time, as many of the modern hybrids use an internal combustion engine at one end and an independent electric drive at the other with computer-controlled torque distribution between the two.

The Hardaker Mk 2

The car is now more conventional in that it is a mid-engined, rear wheel drive two-seater. It looks nothing like the Mk 1 because of the different upper body but under this it includes all the satisfactory elements of the twin engined Mk 1. The chassis retains an aluminium floor with the steel tube front and rear bulkheads with two standard production car front sub-frames, complete with suspension, attached at front and rear. The deep aluminium side sills have been replaced but smaller rectangular section tubular steel sills of equivalent strength and stiffness to improve access through the doors and side impact protection.

The Mini front sub-frame has been replaced by the similar Austin Metro sub-frame in order to convert to "Hydrogas" suspension which is much better for road use than the Mini rubber cones. A Peugeot 205 front cross member is fitted in place of the Mini sub-frame at the rear with the suspension modified to operate "Hydrogas" suspension units. A Peugeot 205 diesel engine and gearbox and a triangular fuel tank are fitted behind the 2 seats separated from the driving compartment by a fire wall bulkhead.

The replacement upper body was made by cutting down a salvaged "Mini Minus" fibreglass kit car body, with suitable modifications. The car therefore looks a bit like an old Mini but is 4 inches lower. Only the window glass, the boot lid and the doors are Mini parts, and the doors have 4 inches cut off the bottom. The mid-engine allows the front to become a storage space making it a much more useful road car than the Mk1.

Controls are more or less conventional, although I finally solved the problem of connecting the gear lever to the rear side of a rear mounted gearbox, a problem never totally eliminated on the twin engined Mk 1. The answer is that all linkage connections between the chassis and the engine gearbox unit should be by means of rotating shafts or tubes with no push-pull rods. This prevents movement of the engine on its rubber mount from interfering with the gear change.

Access to the mid-engine is provided by a removable panel above the engine or if more access is required, the whole of the top bodywork is hinged at the front and can be raised at the rear by removing 4 bolts without and disconnection of electrical wiring or other connections.

I judge the result to be a success, lightweight, simple, and economical with plenty of torque and good balanced handling with the large back wheels and smaller front. It has been progressively modified and improved and is still in occasional use today, although not as often as in the 1990s when it was my only road car for several years and covered a large mileage without significant problems. Perhaps there are further modifications still to come.

PART 3

Application to a Motorcycle

Having got to grips with the basic principles of failure analysis, (basically, to identify which parts of your machine are **Critical** to safety and making sure that they will never fail in use), this part of the book looks at the various parts of a motorcycle to give a guide to what parts are likely to be **Critical** and how to avoid any likely failures. This is only a general guide as individual designs differ, and very minor parts can be **Critical** in some designs so all new or modified parts need to be considered when building or modifying a motorcycle. I will work from front to back with a few notes on each major part of a typical motorcycle together with my own experiences and other unfortunate examples where relevant.

1 – Wheels

Any failure of a wheel which either locks the wheel or causes it to collapse will undoubtedly be **Hazardous**.

Apart from possible brake problems, discussed in another section, the most likely cause of wheel locking is heavy contact between the wheel and some static part at the rim or tyre radius. Failure analysis should look at clearances in this area, for example mudguard and mudguard support clearances and clearance between wheel and frame, suspension or fairing parts, checking the steering from lock to lock at the full bump and full extension of the suspension movement and all points between the two. I suggest that the minimum radial clearance should be at least half an inch on a classic bike to allow for any deflections under load and for tyre growth at speed. With no tyre growth effects across the width on the tyre, about half this clearance should suffice for minimum clearance to the side of the tyre and wheel rim. Modern bikes with larger tyre sizes and higher speeds will require more radial clearance and the best way to set a minimum clearance would be to look at other production bikes with similar performance and the same tyre size. It is also a good plan, particularly on an off road bike, to arrange for the minimum clearance to the tyre to be at the rear end of the mud guard, or similarly positioned part, so that any large object which may be carried round with the tyre will not be carried into a wedge shaped decreasing clearance as it passes between tyre and mudguard, There is a very small chance of this happening but it would be **Hazardous** if it did and this sort of change costs very little in either weight or cash.

Investigation of wheel and tyre clearance at full bump is easier with the springs removed from the suspension as it is difficult to hold the springs compressed while turning the steering and simultaneously checking the clearance.

The other main point to look at is the security of static parts close to the wheel which could become loose and rub on the wheel.

I know of at least 2 cases, both with telescopic forks, where this could have been a real problem if we had not caught it at an early stage. As previously mentioned, I modified a Vincent Rapide many years ago to fit Norton telescopic forks (see photos 6, 7 and 11) and had to cut the rear of the front mudguard away to provide sufficient clearance to the engine at full bump. There appeared to be plenty of clearance when static but at full bump there was zero clearance between the mudguard and the engine, the exhaust, and the forward extended magneto cover at some steering angles. Not good for keeping road dirt and water off the engine but better than the possibility of the wheel or steering locking. The clearance between the wheel and the engine looks very similar to the original suspension system with the suspension unloaded, see Photos 27, but the difference is that the standard suspension moves the wheel almost vertically as it is compressed while telescopic forks move the wheel backwards along the line of the forks. In the end this change, together with some changes to the rear suspension, worked out very well making the somewhat vintage handling of the Vincent equal to that of the latest sports bikes of the day. In terms of miles covered and years of use, this was the most successful of my specials although later ones were more interesting to the mechanically minded.

Standard Vincent front end.

My modified front end

The Elgi Vincent front end (steering head moved forward). (This immaculate restored Elgi was displayed at the Jurby Festival in the IOM. Owner unknown but he has made a very fine job of it.)

Photos 27 – Wheel to engine clearances.

My Vincent modification was many years before the Elgi-Vincent (also shown in Photos 27) which had a similar suspension arrangement but also included a new design of top frame member which moved the steering head forward eliminating the clearance problem.

There is always more than one way to approach a modification, so I have included for interest, before and after photographs of a superb Vincent Black Shadow owned and modified by Paul Adams of the Vincent Owners Club. This one used 1970s Honda telescopic forks and wheels, and the combination of fork length and the smaller wheel diameter allowed sufficient clearance for a full-length front mudguard. All things have advantages and disadvantages, but I believe that this solution would make my preferred steering geometry, with a steeper steering head angle and reduced trail relative to the standard machine, difficult to achieve.

The Vincent Black Shadow was very similar to my Vincent Rapide apart from a few additional performance modifications.

Photo 28 below shows the modified version and Photo 7 shows the standard bike.

Photo 28 – The Paul Adams Vincent Black Shadow.
(photo reproduced courtesy of Paul Adams.)

There is a lot (perhaps too much) about Vincents in this book and they will be, to many younger readers, just some sort of old bike not worth thinking about, but in their time, Vincent were one of the very few manufacturers who combined genuine new thinking with sound engineering, at a time when most manufacturers were just updating traditional designs.

Their 1000 cc engine was also one of the very few motorcycle engines with an aero engine derivative, the "Picador", which was capable of 65 bhp in continuous operation on Avgas aviation fuel (a lot in those days when mechanical and ignition system limitations resulted in much lower maximum engine speeds than are common today). They were intended for early pilotless drones. The Norton rotary is the only other motorcycle engine I know of with an aero derivative developed by the manufacturer, but no doubt there were a few more and there have been some "aftermarket" aero conversions.

Unfortunately, all this comes at a cost which must be recouped from sales and this did not prove possible in the long run. However, I always liked them and wish I still had mine.

Moving away from Vincents, the other case of a front wheel clearance issue was more recent, on a bike built by a local motorcyclist, I believe largely as a show bike but road legal and capable of being used on the road. A turbocharger arrangement, designed for another installation, had been fitted in front of the crank case on a relatively modern 4-cylinder Suzuki. In this case, contact with the turbocharger casing completely locked the wheel at full bump although clearances looked to be sufficient at the normal static ride height. As moving the turbo was classed as "too difficult", the builder's initial solution was to make new fork yokes moving the fork legs forward. This cleared the foul but would obviously reduce trail and I have not yet heard of the effects on handling. I hope one ***Hazard*** *has not been corrected by introducing a potential instability, which could itself be* ***Hazardous,*** *but he has not found it to be a problem to date.*

The possibility of wheel collapse may be seen as less likely than locking due to contact with a static object but as it would definitely be **Hazardous,** we should consider the possibility.

A wire spoked wheel has an advantage here over a lightweight cast wheel in that failure of a few spokes would reduce the stiffness of the wheel and is likely to be readily detectable when riding or cleaning long before total collapse occurred.

The most likely problem with a cast or forged wheel is fatigue cracking, most likely at the junction between the spokes and the rim and this area should be carefully inspected if using a second-hand wheel. You could argue that a greater number of spokes will make the wheel safer from a cracking point of view, but I prefer a small number of large spokes (as the ZXR400, Photo 22) as they are much easier to clean and inspect regularly.

Out of interest, the difference between a forged and a cast wheel is in the method of manufacture. A cast wheel is manufactured by pouring molten metal into a mould to produce the basic shape of the wheel, then finish machining the areas which need to be dimensionally accurate such as the bearings locations and brake disc and tyre locations. A forged wheel is produced from a piece of metal heated until soft but not melted. The basic shape of the wheel is then produced by, in effect, squeezing the metal blank between shaped formers and is finished by machining where necessary as with a cast wheel. The result is that a cast wheel will be cheaper to produce but not as strong as a forged wheel, all other things being equal. Also, material defects such as areas of porosity or inclusions, where small particles of grit or similar substances have contaminated the molten metal, are more likely in a casting and could provide origins for fatigue cracking.

I would always suggest that, wherever possible, you use a production wheel from a bike in general use, on a road-going special, to benefit from the evidence of satisfactory operation in service. Special lightweight wheels may have theoretical advantages in reduced un-sprung weight and gyroscopic forces, but their reliability is unproven, and the effects of failure could be somewhat **Hazardous**.

Wheel failures are rare, but they can happen. Robert Dunlop had an unfortunate accident shortly after the jump at Ballaugh Bridge in the 1994 Formula 1 TT, which was widely reported in the press to be as a result of failure of a lightweight rear wheel. I never heard what the official cause of failure was, but the pictures published at the time suggest that the most likely cause would be development of fatigue cracking during previous running to the point where a high single load, perhaps from landing the jump, fractured the remaining metal.

The trend towards carbon fibre wheels as a further weight reduction brings its own problems as the integrity of a carbon fibre component depends entirely on the level of quality control in its manufacture. Manufacturing defects such as low matrix material strength or low levels of adhesion between laminates of material or errors in the internal pattern of laminates are difficult to detect on the finished product. Ultrasonic testing is the best inspection method but perhaps not infallible. The only thing a special builder can do if he intends to use carbon fibre wheels is to make sure he knows and trusts the manufacturer, and if the wheels are second-hand, knows their history.

For interest, composite materials such as carbon fire or glass fibre composites are composed of high strength, but flexible fibres held in position by being embedded in a relatively low strength matrix material, in our case normally known as "the resin." They fall into basically two types, random fibre, in which the fibres are not aligned but are randomly distributed in all directions, and aligned fibres where the bulk of the fibres are aligned parallel to each other in particular directions. This results in the random fibre material having roughly the same strength in all directions but not using the full strength of all the fibres. Aligned fibres, either unidirectional with all the fibres arranged parallel to each other in one direction, or woven cloth, with half of the fibres aligned in one direction and the other half at 90° to that direction, have much greater strength than random fibre in the directions of fibre alignment but can introduce planes of lower strength between individual fibre laminates where very few fibres cross the plane of matrix material. Also aligned fibre parts are much less strong in compression than they are in tension because the fibres if unsupported by the resin would buckle at very low loads. Compressive strength therefore depends more on the strength of the resin keeping the fibres aligned than on the strength of the fibres.

It is therefore obvious that the internal design of a composite **Critical part** needs to be very carefully arranged such that aligned fibres are positioned in the directions of the major stress and that any planes of lower strength are not subjected of significant stresses. It is equally obvious that a good design is of no use if the finished product does not conform to the design, in our case if the internal layout of the fibre structure is not as designed.

The other factor in manufacture of high-quality composite parts is the heat curing process which hardens the matrix material. Careful control of the time and temperature of curing is required together with attention to the means of compaction to ensure that there are no air or vacuum filled voids inside the component, and to avoid distortion during curing.

Unfortunately, it is often impossible to check the internal fibre layout or the quality of the manufacturing process on the finished product, so we are entirely dependent on the care taken during manufacture for the strength and durability of the product. I have seen destructive tests on test samples for prototype composite aero engine components, produced either by cutting the part into sections or heating in a vacuum to remove the resin without disturbing the fibre layout, where the fibre layout had distorted considerably during the curing process.

Sudden tyre failure could obviously have **Hazardous** effects but using correct tyres approved for the speed and weight of the machine will give the same level of safety as any other production machine so need not concern us from a failure analysis point of view.

I always use a proprietary sealant in the tyres as an extra precaution and have seen no ill effects from this, not even significant out of balance on a motorcycle wheels, (although I tried it once in a car tyre and was unable to get a satisfactory consistent balance). The sealant is a good protection against a sudden deflation at speed, which could be ***Hazardous,*** *as it will prevent or slow deflation in most punctures caused by nails or similar small objects. It could be called a* ***"Safety System"*** *in that it does nothing for the vast majority of its life and could be called on to act instantaneously on the one occasion when it was needed. (A better* ***safety system*** *than most mechanical or electronic* ***safety systems*** *as it does not need frequent checks to ensure that it is still working.)*

Wheel bearings, provided that they are a suitable size and type are generally reliable and the only deterioration likely to be seen is gradual wear which would be detected by the rider as movement at the wheel rim before any **Hazardous** effects could occur.

The possibility of seizure of a wheel bearing, which could be **Hazardous**, can generally be discounted on the basis of experience. In most cases, if using a suitable production wheel with the standard bearings, reliability should be as good as on the donor production machine.

The only exception I can think of is perhaps a hub centre steering design which may require large diameter non-standard wheel bearings and would need careful consideration of failure analysis aspects.

As stated above, the only likely failure mode of a motorcycle wheel bearing is wear, but for interest, there are other failure modes of rolling element bearings which are worth mentioning although unlikely to be **Hazardous** in our case even if the deterioration is severe.

- Track fatigue can occur in highly loaded bearings after long period of use. It is the result of repeated compression and relaxing of the track surface as each ball or roller passes over it. This loading can cause sub-surface cracks to form parallel to the track surface which eventually progress to allow small flakes of the track surface to be released. Overloading by use of a bearing too small for the job will accelerate this process. Fatigue damage is a slow process and if it did occur on a motorcycle, it would probably be detected by a rough feel when rotated before there is any likelihood of total failure.
- Corrosion is unlikely on a lubricated bearing, but neglect can result in track and ball corrosion which will accelerate track fatigue.
- Extreme impact loads can cause indentation of the tracks. This is most likely on a static bearing, such as a steering head bearing, which may have been subject to many "heavy landings", rather than on a continuously rotating wheel bearing. Again, it should be detected by feel before total failure occurred.
- Lack of lubrication on a wheel bearing would generally just increase the rate of wear at the sort of bearing loadings seen on a motorcycle.
- In more extreme cases of heavy loading and high temperature, such as the main thrust bearings in an aircraft gas turbine, failure of the oil supply can result in total destruction of the bearing. These bearings transfer the high forward or rearward loads generated in the compressor and turbine rotors to the main engine structure and thus to the engine mounts and rely on oil flow to carry away heat generated by friction. Loss of this cooling effect can result in a very rapid increase of temperature in the bearing. I have seen bearings where sufficient heat has been generated to melt the surface of the bearing balls and track. This is an extreme case where the power of the turbine is so high that the increased friction in the failing bearing will not significantly slow the turbine and the failure would only be detected when the resulting increase in clearance in the bearing allowed the whole rotor to move out of position generating excessive vibration of the whole engine on its mountings as a result of contact between the rotor and static parts and rotor out of balance.
- Skidding of the balls or rollers can occur on bearings rotating with no load applied and can cause track and ball/roller wear. The cause of skidding, where the ball/roller does not rotate at its natural rolling speed on the track, is friction at the only point where sliding contact occurs in a rolling element bearing between the ball or roller and the cage. Low radial or axial load on the bearing can in some cases result in insufficient driving force between the ball/roller and the track to keep the ball/roller rolling particularly during acceleration or deceleration.

The above may have wandered off the point again, but perhaps may be of interest to some readers.

The wheel spindle is another **Critical part** which not only supports the wheel but also, in many designs, including the normal telescopic forks, keeps both fork legs moving together and thus keeps the wheel upright relative to the rest of the bike. It is worth remembering that bump loads are still vertically upwards even when the bike is banked over in a corner so the wheel is loaded sideways by bumps tending to compress the "top" fork leg and extend the "lower" leg. The wheel spindle keeping the legs moving together is thus required to resist bending into an S shape due to vertically upward bump loads in addition to the shear loading resulting from the weight of the bike and the G-forces resulting from turning the bike.

Again, our best answer on this is to ensure that the spindle is at least as strong, in both bending and shear strength and in fatigue strength, as a similar arrangement on an equivalent production machine of similar weight, both with respect to dimensions and material strength and quality.

Sketch 12.1 below illustrates the forces on a wheel during cornering.

The sketch shows a lean angle of 30° from the vertical but it is easy to see that the effect of bump loads would be much more pronounced at higher lean angles.

The effect of a rider moving his body weight by hanging off and leaning his body further into the corner is the opposite of a bump load, introducing bending into the wheel in the opposite direction but this effect is not likely to be very great.

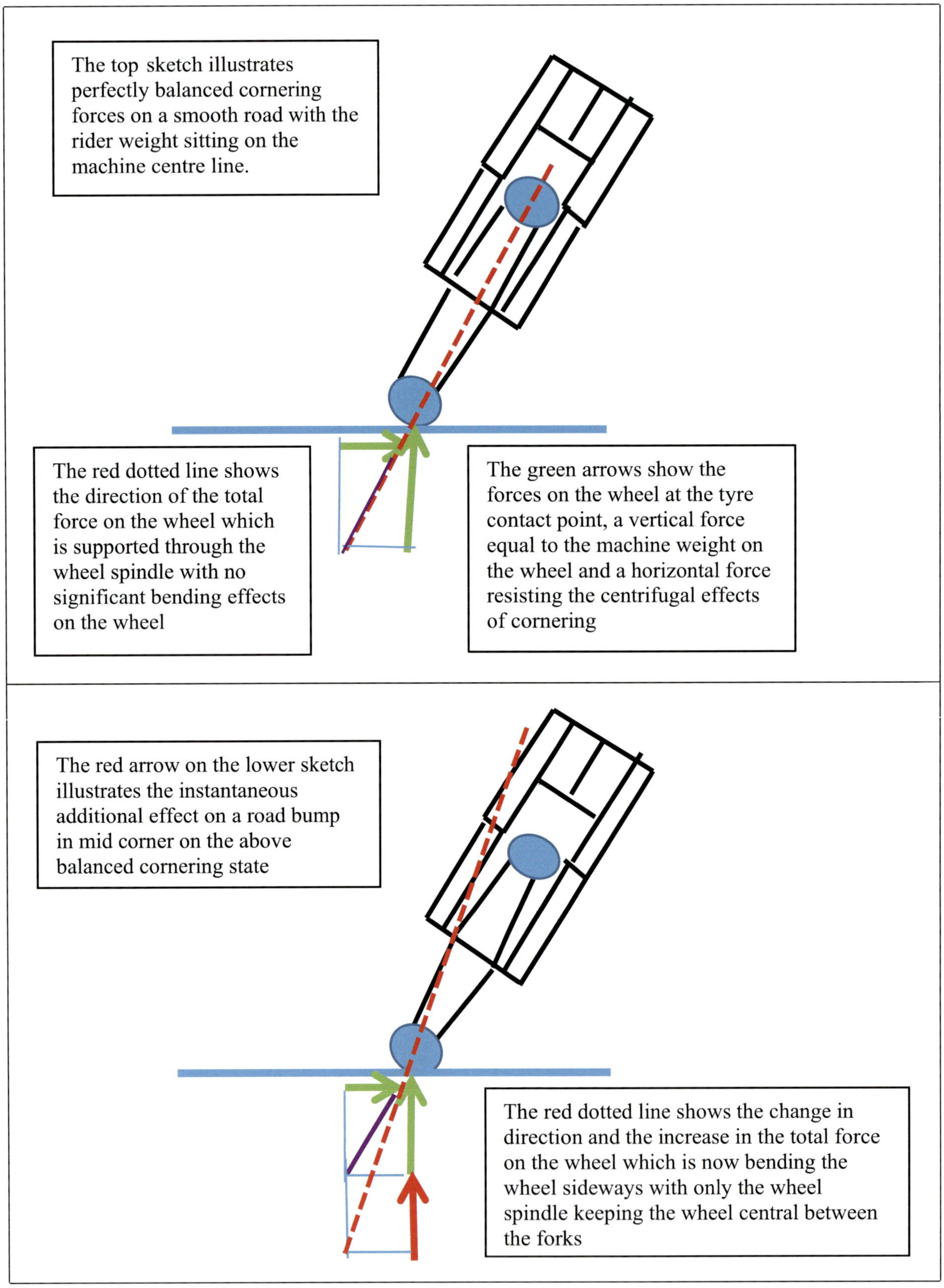

Sketch 12.1 – Showing the loads on a wheel during cornering.

The following check list is included as a guide to the items which require some thought to avoid possible **Hazards** although these will vary somewhat dependent on the design.

Wheels		
Type of failure & possible effects	**Possible causes**	**Typical justifications for a satisfactory low probability of failure**
Wheel locking due to road or loose debris. Effect: Hazardous Probable loss of control	Road debris wedged between wheel and static parts	Is minimum clearance satisfactory? Are there any decreasing clearance wedges between the wheel and the static parts? If so, consider increasing clearance.
Wheel locking due to contact with static parts. Effect: As above	Attached parts close to wheel becoming loose or moving	Are there any likely parts which could become loose or move? If so, why will they remain secure? Is there anything clamped to a fork leg or similar part in a position which could rotate in the clamp turning it into the wheel? If so, what stops it rotating? (loose clamp bolts are not unknown)
Wheel clearance at full bump and/or full lock Effect: As above	Contact with static parts	Have you checked, preferably with the springs removed, to ensure that all possible steering and suspension positions have been covered? Is there enough clearance at the closest point to allow for tyre growth and some compression of bump stops or rubber bushes, if fitted? If not, consider how the minimum clearance could be increased.
Wheel collapse - Cast or forged wheels Effect: As above	Cracking	Are the wheels standard parts of a production bike of similar or greater weight? If the wheels are not new, do you know their history? For example, not from a stunt bike or a front-end crash. If they are suspect, suggest careful inspection for cracking, particularly in the spokes, and a check on the rim for running true, or look for another wheel.
Wheel spindle loose Effect: Reduced rigidity of fork assembly allowing wheel to lean relative to forks. Deterioration in handling. Possible contact between tyre and fork parts during cornering. Not likely to be immediately Hazardous but should be corrected as soon as detected.	Inadequate tightening or locking of nuts.	Is there a secure locking method for the wheel spindle nuts? A split pin is best in my opinion, but other positive locking methods can be equally effective.

Wheel spindle failure Effect: Hazardous Probable loss of control	Unsatisfactory design or poor material. Possible ultimate effect of long-term running with loose nuts.	Is the spindle a standard part designed for the wheel? If not, is it similar in both dimensions and material strength to the standard spindle? If not, consider the possibility of stronger spindle. Are the wheel spindle nuts tight and secured as above? If not, improve as necessary.
Tyre failure Effect: Possible loss of control as a result of rapid deflation at speed, which would be Hazardous.	Unsuitable tyre choice or unsatisfactory fitting Damage from road debris.	Are the tyres the correct weight and speed rating for the intended use? Are the rim widths suited to the tyre choice and the tyre fitting positions on the rim undamaged? Have the direction of rotation and front/rear tyre markings been adhered to? Are the tyres fitted correctly with the circumferential line parallel to the rim all the way round and at the same distance from the rim on both sides of the wheel? If an inner tube is used, is it fitted correctly, not trapped under the bead and with the valve projecting at 90° to the rim before the valve ring nut is fitted? Are the selected tyre pressures suited to the weight on wheel? If not, correct as necessary. If the above points are all satisfactory, the probability of damage from road debris, which could cause sudden deflation and therefore could be Hazardous, will be the same as in any other use of the tyre and therefore acceptable for our purposes. Have you considered using a permanent sealer in the tyre which would prevent sudden deflation in all cases of penetration by nails or other similar road debris, leaving the very unlikely case of a very large split as the only Hazard?

2 – Front Suspension and Steering

There are many possible types of front suspension we could use on our motorcycle, and each has its own **Critical parts**, telescopic, girder, leading link, trailing link, BMW type tele lever and the various hub centre systems, to name but a few.

The easiest type to use, from a failure analysis point of view, is a standard telescopic fork from a production machine of similar weight and performance. The sliders, main fork tubes, yokes and steering head stem are all **Critical parts** but by using a tried and tested production set up we can ensure an equivalent level of safety and a satisfactory low probability of failure.

In practice, the suspension and handling which can be achieved with a good telescopic fork arrangement is hard to better, but part of the appeal of building a one-off motorcycle is to try something different. Every type has its own failure analysis problems and its own set of **Critical parts**, the failure of any of which would result inevitable loss of control.

At the other end of the scale of difficulty is a one-off design for a hub centre system where there are inevitably more parts and pivot points than with a telescopic fork and almost all the parts would be unique. Much thought would be needed to design and make a hub centre steering system to minimise the number of potentially **Hazardous** failure types and to ensure a satisfactory low probability of those which cannot be eliminated by design.

I attempted the design of such a system at one stage and made some parts but, for various reasons, it was never completed, and I went for a leading link design instead. Perhaps one day I will try again as there are theoretical advantages, but detailed design is difficult, particularly in minimising the width of the necessary horizontal wheel hub rear locating links pivoted on the frame while still allowing sufficient steering angles, and avoiding stick/slip friction or free play in the steering linkage between the handlebars and the wheel while retaining absolute consistency in wheel movement relative to handlebar movement.

The following points need to be considered for all types.

STRENGTH OF CRITICAL PARTS

In the case of systems using a normal steering head, the stiffness of the fork legs is an important design consideration because of the large lever arm from the point where loads are applied, at the tyre contact point through the wheel spindle, to the point where they are reacted through the steering head. A sufficiently stiff unique fork leg, comparable with the equivalent telescopic fork, is likely to have equivalent strength and fatigue life if it has similar dimensions and is made from suitable material of satisfactory quality. Thus, we would achieve a similar low probability of failure to a production set up even if our forks are very different in design.

Similarly, with the steering stem and yokes, and all **Critical Parts**, similarity with a comparable production set up is the easiest way to ensure a low probability of failure.

All the types of suspension, other than telescopic types, usually incorporate various extra structural parts and pivoted links which are generally all **Critical Parts** and are likely to have no direct production equivalent. The following points are common to all types and the effects of possible failure of each should be considered in each case.

CLEARANCES

The question of clearances keeps coming up, but it is worth mentioning again as it is so easy to get it wrong by just a visual check with the machine static and unloaded. The full range of steering angle and suspension movement should be reviewed to ensure that, at all points, there is good clearance between static parts and steering parts and between the wheel and all other parts, to allow for possible deflection, debris trapping and tyre growth, etc.

SUSPENSION COLLAPSE

Suspension collapse, for instance due to failure of the mounting of a single suspension unit is not impossible but can usually be made **non-Hazardous** by arranging an independent bump stop contacted before the wheel or the steering is locked. If two suspension units are used, the problem does not arise as simultaneous failure of both units, or their mountings, is unlikely and the second unit would generally prevent total collapse.

FAILURES WHICH COULD LOCK THE STEERING

Steering pivot wear or seizure is unlikely to be a problem if standard production parts or their equivalent are used and are correctly adjusted for clearance/pre-load, on the basis of the large amount of satisfactory road-going experience with similar taper roller or ball thrust bearing systems.

In any type of unusual front end, the steering pivot would normally still use ball or roller bearings which are not prone to sudden seizure or rapid wear. Gradual increasing stiffness, roughness or looseness would therefore normally be detected by feel before it became **Hazardous**. I have found that the first symptom of increasing friction on a steering head is a low-speed weave probably due to the stick-slip nature of the static friction prior to the initial movement. You get a similar effect with the old friction type steering dampers used on some classic bikes or if the steering head bearings are adjusted with too much pre-load.

This potential for stick-slip static friction is the reason why plain bearings, including car type ball joints, are never totally successful as heavily loaded motorcycle steering pivot bearings, however attractive they may seem to the special builder from a space point of view.

The short lived Norton Nemesis project in the 1980s was said to have plain bearings in the steering head and press reports at the time commented on a low speed weave, which was most likely to be due to the stick slip friction effect.

Having established that sudden steering head bearing failure is not likely to occur, we should look at the items around the steering head area to see if failure of any of these could lock the steering.

Relatively minor failures, such as any of the following, could lock or restrict the steering and thus become **Hazardous**. The probability of each should be considered.

- Failure of a steering damper mounting (one loose bolt), if the unit could move into a position where it interfered with the steering.
- Breakage of a fairing mounting which allowed the fairing to foul the handlebars.
- Rotation of the handlebar relative to the forks, particularly in the case of clip-on handlebars where I think an anti-rotation feature of some sort is desirable even though most classic racers just clamped the half handle bars to the fork legs.
- Failure of a non-standard steering lock or the possibility of riding off with it in an unsafe position which could engage at any time. Engagement should be impossible, whatever the position of the ignition key or equivalent, unless the steering has been turned as far as possible from the centre position.
- Failure of the attachment of handlebar cables or wiring to the frame which may then move to a position which restricts handlebar movement, e.g. trapped in the steering lock stops or fouling the fairing, etc.

Any of the above failures could suddenly become **Hazardous** and it is too late to discover the problem when you are riding down the road. All parts which could possibly fail should therefore be reviewed for possible **Hazardous** effects and suitable precautions taken against all the **Hazards** discovered.

My own experience of "funny front ends" is mainly with leading link arrangements which I used in two specials. The supercharged Velocette Valiant in the 1960s (photos 4 and 13) and a supercharged Triumph Tiger 100 based special in the 1970s (see photos 14 and 15). Both worked well compared with the contemporary telescopic fork machines, particularly on braking where the built-in 80% anti-dive geometry, which did not interfere with the suspension movement, was a real advantage. The remaining 20% dive proved to be sufficient to give the desirable feel for the level of braking applied. They both used less trail in the static position than a typical telescopic arrangement, but the links were arranged such that the wheel moved backwards relative to the line through the steering head bearings as the suspension deflected upwards giving what may be considered as a normal amount of trail [4 inches, (approx.100mm)] at full bump. My experience of this arrangement is that it gave sensitive steering and was not prone to any detectable steering instability with no steering damper fitted, even when provoked. Failure analysis-based precautions built into the design included:

- *The two leading links and the front wheel spindle were one-off parts designed to be stronger than the fork legs which were taken from a production fork as used on a motocross machine of similar weight. Failure of these parts would still be **Hazardous**, but a satisfactory probability of failure, no worse than the production standard fork tubes, could be justified.*
- *The link pivot points at the bottom of the fork legs were machined from steel and included a spigot which fitted inside the fork tubes, secured by cross pins and brazed, thus avoiding a critical weld position where the available weld quality could not be guaranteed. Again, failure of any one of these parts would remain **Hazardous** but a probability of failure, no worse than the production standard fork tubes to which they were attached, could be justified by the fact that the pivot fittings were stronger than fork tubes.*
- *Twin suspension units where used, one on each link with the two leading links connected together behind the wheel. The connecting U-shaped part was sufficiently stiff such that the removal of one suspension unit did not significantly affect wheel alignment and just resulted in a lower ride height. This design relieves a large proportion of the bending loads on the front wheel spindle when crossing road bumps banked over by ensuring that the two leading links always move together and also means that failure of one suspension unit or unit mounting would not be **Hazardous**.*
- *One of the main advantages of non-telescopic front ends is the ease with which anti-dive during braking can be arranged by suitable angling of the brake torque reaction links. However, this introduces additional links, with pivoted end fittings etc., and thus more possible failure points compared with the usual solid bolting attaching the brake callipers to telescopic fork sliders.*

*The brake torque reaction links and pivot bearings on my machines were also one-off parts but were designed to be stronger than the standard attaching bolt to the standard (Honda) drum brake back-plate. In addition, to avoid a possible **Hazard** if the torque link became disconnected, a projection was arranged on the brake back-plate which would contact one of the suspension leading links and stop rotation of the back-plate with the wheel after a small angular movement. The brake operating cable was long enough so that, even when operating on this stop with the torque link failed, there was no tension on the cable which would have locked the brake on.*

Thus, by small design changes there were no ***Critical parts*** *in the brake torque reaction system. Without the stop, the torque arm and its mounting points at both ends would all have been* ***Critical parts****, the failure of any of which would have been* ***Hazardous****.*

As another distraction from the subject of this book, the mention again of bump loads while banked over made me think about the much-discussed phenomenon of chatter, which seemed to plague Moto GP bikes a few years ago, but was not much of a problem elsewhere even on production based racers. This seems to be generally agreed that the solution to this problem is to change the torsional flexibility of the frame which presumably allows some sideways deflection of the front wheel under vertical bump loads when at extreme angles of lean.

It seems to me that the basic cause of chatter is likely to be the fact that bump loads, however small, remain vertical and thus are not aligned with the direction of suspension travel when at high angles of lean. The only "suspension movement" remaining in these conditions is flexing of the tyre walls and any torsional flexing of the frame and forks. On a Moto GP bike everything is against you in this respect, with very high lean angles and stiff tyre walls, wheels and forks and relatively low weight and inertia of the whole bike to provide some mass for the suspension spring to react against. In these circumstances, even very small bumps will result in rapid variations in the tyre to road downforce and thus rapid cyclic variations in grip. Torsional flexing of the frame, in effect letting the front wheel move sideways relative to the bike centre line, is perhaps not the best way to provide some sideways compliance for the front wheel as this "spring" is completely undamped and the whole of the front end is effectively un-sprung weight. You could imagine that some sort of rubber bush at each end of the wheel spindle to provide the sideways compliance between the wheel and the forks together with a very stiff frame may a better solution. This would reduce the effective un-sprung weight if the wheel was deflected sideways by a bump at full lean, provide some internal damping in the rubber and allow quick easy changes in the spring rate, by replacing the rubber blocks with another set with different stiffness, during setting up. Also, constant wheel steering angle relative to the handlebar angle would be maintained avoiding any possibility that frame flexing could result in unintended rider inputs to the handlebars, and thus to steering angle.

Such a system would need clever detail design to allow the brake callipers to maintain position relative to the discs, but this should not be impossible. Sketch 12.2 below shows the basic idea.

Just a theory but might be worth a try.

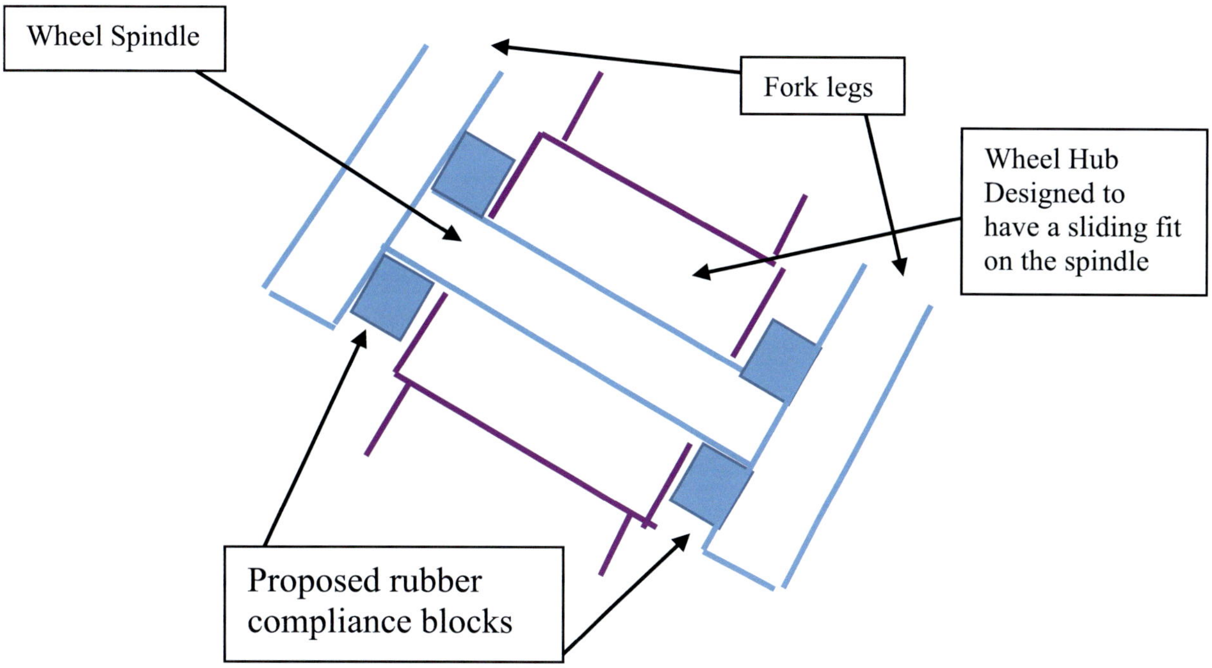

Sketch 12.2 – Possible method of introducing sideways flexibility to the wheel?

Back to the main subject.

The following check list is included as a guide to the items which require some thought to avoid possible **Hazards**. Everything which is not directly taken from an equivalent production machine should be reviewed for possible failure effect, however unlikely.

Front suspension and steering		
Type of failure & possible effects	**Possible causes**	**Typical justifications for a satisfactory low probability of failure**
Structural failure Metal parts Effect: Probable suspension collapse and loss of control	Insufficient strength for maximum loads applied.	Are the parts taken from, or stronger than, those from a comparable production machine? If not, are there satisfactory calculated strength reserves at Critical positions relative to best estimate of loads applied? or Can we conclude that failure at the position in question is not Hazardous due to an alternative load path or that there could be no chance of interference with the steering or braking?
	Poor material quality	Is the source of the material and material quality standard known for all one-off parts? If not, what sort of inspection/testing has been carried out?
	Unsatisfactory weld quality	Were the welding methods and materials suitable for this joint? What sort of inspection of welds at Critical positions has been carried out?

Structural failure Composite parts	Failure of Composite materials (e.g. carbon fibre) Undetected manufacturing defects Impact damage from road debris	I think the best advice to anyone considering using composite Critical parts in a front suspension system is "Forget it" It is always possible to get the equivalent strength and stiffness with metal at the expense of a small weight increase which will be insignificant on a road bike as the front suspension is only a very small proportion of the overall weight of the bike. For the reasons described in Section 1 of Part 3 covering **Wheels**, it would be impossible to guarantee the manufacturing quality of a unique carbon fibre composite component with sufficient confidence to justify the necessary low probability of failure of a Critical part. In general, apart from wheels where reduced gyroscopic effects may be a noticeable improvement on a fast road machine, there would seem to be no advantage to be gained from the use of carbon fibre Critical parts on a road bike. The only motorcycle use of composite Critical front suspension parts I know of, is on the Britten racers but there may be a few others. Formula 1 cars often use carbon composite struts in suspension systems but even with their expertise in this area, they are not immune to the occasional unexpected failure with some vastly expensive wreckage the result. If the appearance of carbon fibre is important to the visual appearance of the bike, suggest that unstressed carbon fibre covers over metal load carrying parts would be the way to go.
Clearances Effect: Possible wheel locking which may be just instantaneous and immediately released but could, in the worst case, result in inability to turn the steering and loss of control.	Wheel clearances to frame parts.	Is there sufficient clearance between the wheel and all adjacent non-rotating parts over the full range of steering lock to lock and the full range of suspension movement, taking into account clearance required for tyre growth at speed? If not, improve the design
Bump stops Effect: Lack of effective bump stops could permanently lock wheel in the event of suspension collapse causing loss of control.	Suspension collapse.	What stops the wheel from contacting stationary parts if a suspension unit failed or became detached? If the answer is nothing, consider adding a bump stop of last resort.
Interference with steering movement	Insufficient steering clearances	Is there anything which could interfere with handlebar movement? e.g. in the event of minor deflection of fairing parts or interference from cables or wiring. If any areas are suspect, consider how they could be improved.

Effect: Could cause loss of control	Mechanical failures	What would be the effect of a detached, bent or damaged steering damper or similar part? If it could lock the steering, consider how the design could be improved. How is the unintended engagement of the steering lock guarded against? *(The usual safety arrangements are for the key mechanism to prevent engagement when the ignition is in the on position and for the mechanical lock to be arranged so that it is only possible to engage it with the handlebars turned as far as possible from the straight-ahead position)*
	Rotation of a clip-on type handlebar on the fork tube Inadequate Locking feature	What prevents rotation of the bars on the fork tubes? *Older type clip- on bars often rely on friction from the tightening of the clamp bolts as the only means of preventing rotation of the bar on the fork tube. This method has a good safety record but from a failure analysis point of view it is better to have some additional positive mechanical locking*

3 – Frame

The frame is the main structural part of most motorcycles to which all other parts are attached, its most important job being to connect the steering head to the rear suspension mounting points and to maintain the alignment of these two ends of the machine under all loading conditions.

In some cases, the engine/gearbox unit takes the place of most of the frame connecting the steering head to the swinging arm pivot and providing attachments for the rear sub-frame and rear suspension mountings and all the smaller attachments such as footrests, fuel tank, fairing etc. This is a very good idea in general as the engine/gearbox unit is a very stiff component, much stronger than most frames, and the design also saves some weight, but not the whole frame weight as the necessary additional attachments such as the steering head and local strengthening on the engine/gearbox unit all add some weight. There is also the downside in that it removes any possibility of building some flexibility into the engine mountings and thus inevitably transmits more engine vibration to the rider. Although a good idea, it is normally outside the scope of the individual special builder because it is likely to require new engine castings unless a donor engine unit of that design is used.

The normal frame of steel tubes or cast and fabricated aluminium sections welded together is obviously a **Critical** part in respect to its main function of holding the steering head, engine and rear suspension in their correct alignment, but many of the possible failures are not **Critical** and would not have **Hazardous** consequences, for instance a fractured fairing attachment point is unlikely to have effects more serious than a flapping fairing because of the numerous other fairing attachment points. The main potentially **Hazardous** failures of a frame are all concerned with maintaining correct location of the wheels, unhindered wheel rotation, no interference with operation of the steering or brakes and support for the engine and gearbox. With this in mind, it is easy to pick out the parts of the frame where a **Hazardous** failure could occur and pay particular attention to each. As with all the other parts, the basic method is first to see if any of the **Hazards** can be removed by changes to the design, (a small additional bracket or extra bolt position can sometimes totally eliminate a **Hazard**) and secondly to justify a probability of failure of the remaining **Hazards** which is no worse than a similar successful production design.

The use of carbon fibre composites for a structure to replace a metal frame has been much discussed recently with claimed advantages in weight and strength and the ability to create enclosed spaces which could be used as tanks or air-boxes within the structure sound promising. However, as far as the one-off special builder is concerned, the same difficulties as mentioned under front suspension components also apply in that, for the **Critical parts** of the frame/structure such as the attachment of steering head and rear suspension pivot points, it is virtually impossible to guarantee the manufacturing quality and integrity of the one–off composite material structure such that the probability of a **Hazardous** failure is acceptable. As I see it, the only way this could be achieved is to arrange the design such that the **Critical** attachments are supported by a number of different semi-independent load paths to the main body of the structure so that cracking of one load path would be detected before any **Hazardous failure** could occur.

Another possibility would be a part metal, part composite frame with a simple relatively light internal metal tube structure connecting the steering head and the swinging arm pivot to provide the necessary basic strength and accurately position the pivot bearings. An external carbon fibre box section could then be bonded to the tubes to provide torsional stiffness and all the other things the frame needs to have.

I used this principle with some success on the deep side sills of the Mk 1 car chassis (Photo 26.1), which provided a good proportion of the chassis torsional stiffness. It used three steel tubes at the corners as the basic structure and a strong light "Dural" sheet outer skin riveted to the tubes to add the necessary torsional stiffness. There are very few totally new ideas, but better materials improve most things.

One potentially **Hazardous** failure, which has occurred in the fairly recent past on a production machine, which shall be nameless, is collapse of the rear sub-frame such that it dropped the back end of the bike onto the rear tyre, potentially locking the wheel. Most rear sub-frames on mono-shock machines are attached at 4 positions to the rear of the main frame by bolts or welding and are thus cantilevered from the back of the frame carrying the weight of the rider and perhaps a passenger and luggage, a considerable downwards load, particularly when travelling on a bumpy road. There is a recent tendency for new designs to reduce the depth of the rear sub-frame by moving the mountings on each side closer together, presumably for the visual effect of a slim support for the seat, but this increases the loads on the sub-frame attachments and is likely to require an overall increase in weight to achieve the same strength. The traditional twin shock frame, where the rider weight is mainly supported directly by the rear suspension units under the seat, has an advantage here, as the rear sub-frame is subjected to much lower loads, and the total weight of the two suspension units and the sub-frame could easily be lower than a mono-shock design.

Most of my machines have used production frames with minor modifications but I have made my own on two occasions which may be of some interest. The first, influenced by the racing successes of the Moto Guzzi and AerMacchi horizontal singles in the 1950s, I built a horizontal twin (see Photos 14 and 15) using a Triumph pre unit twin mounted on its side with the cylinder head pointing forwards and supercharged using a roots type aircraft cabin blower, surplus from an earlier generation of aircraft and available very cheaply at the time.

Another interesting diversion, well off the subject but prompted by the mention of cabin blowers. These are the air compressors used to increase the air pressure in the aircraft cabin so that the aircraft can fly at higher altitude without subjecting the passengers to the very low outside air pressure.

After several decades when the traditional aircraft engine driven cabin air compressor, from an earlier era, has been replaced by cabin air almost invariably being taken from the compressor sections of gas turbine engines, there is a trend back to the separate "cabin blower", now electrically driven and compressing outside air for cabin pressurisation and ventilation.

*This change has resulted from the need to reduce the possibility of oil fumes from the engine entering the cabin in the event of engine oil leakage into the engine compressor air. The low probability of **Hazardous** levels of cabin air contamination provided the justification for using air from the engine compressors and this remains the case. However, actual events have shown that lower levels of contamination can occur with greater frequency, and this has resulted, eventually, in a trend towards design changes to eliminate the problem. The Boeing 787 Dreamliner is one of the first new aircraft designs to use separate electrically driven cabin air compressors.*

*Back to the bike — the horizontal engine required some interesting modifications to the oil system and air-cooling arrangements, but it all worked fairly well in the end and introduced no new potential **Hazards**. The frame was a single 3 inch (76mm) square mild steel tube with the engine and gearbox suspended below it from a rubber bush near the front and a rigid pivot point at the rear of the gearbox such that there was some up and down compliance in the engine mounting but no sideways compliance. (This sounds heavy, but it was lighter and stiffer than a road-going Norton featherbed frame and the frame tube also formed the oil tank eliminating more weight.)*

*Failure analysis considerations for the frame were looked after by low stress levels, largely bolted construction with no total reliance on welds at **Critical** positions and a secondary support for the engine in the event of rubber mount failure. Sketch 13 shows the principles of* *the arrangement which can be accurately made with only hand tools and with no jigs or high integrity welding required; an important consideration when facilities are limited.*

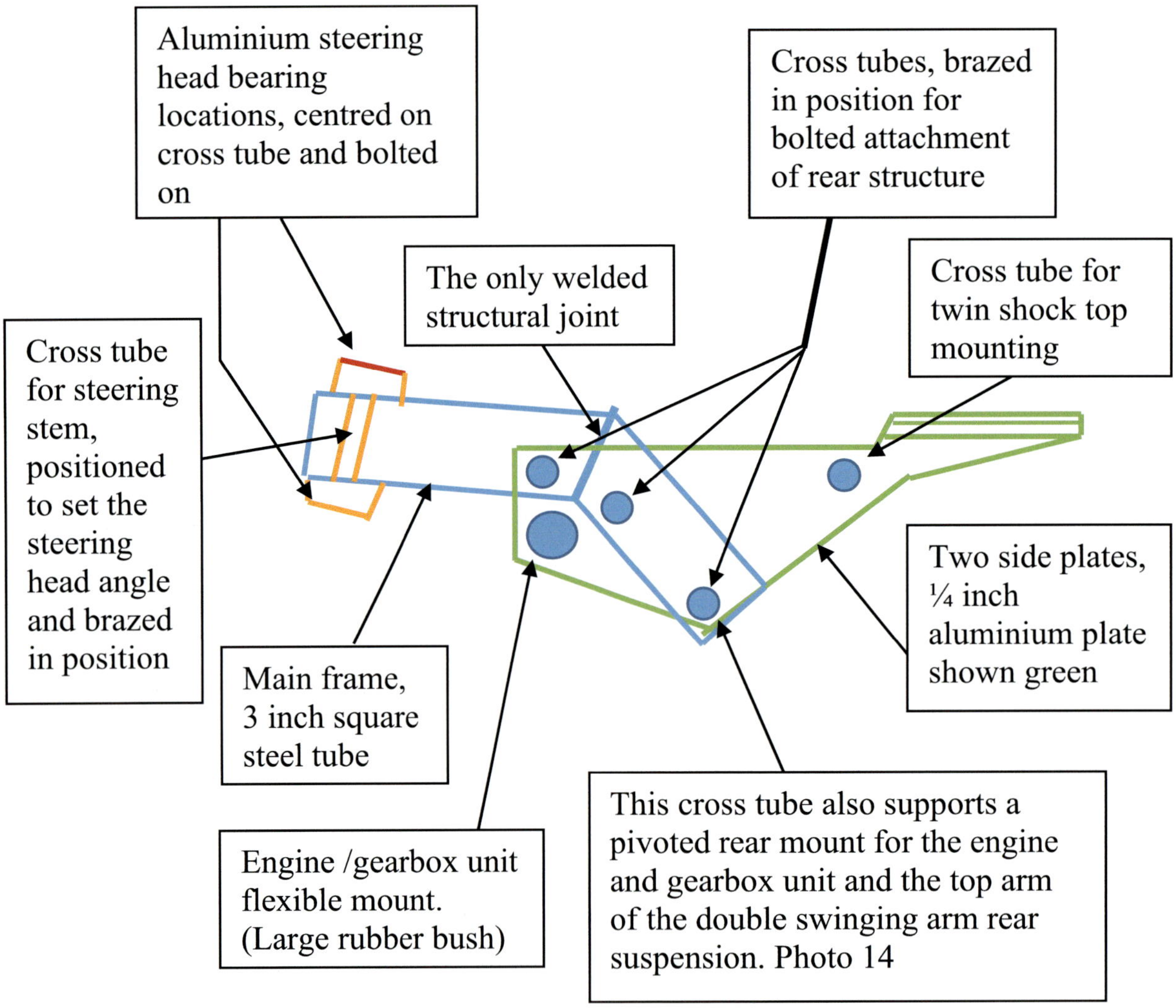

Sketch 13 – The frame for the horizontal Triumph.

The second was a major frame modification for an experimental feet forward machine based on a Yamaha RD400 two-stroke (see Photo 17 and 18). This bike used the Yamaha frame as the front end with a full width box section extension under the low seat to the original Yamaha swinging arm and two fabricated cantilever arms behind the seat to support the tops of the standard twin shocks. In this case, failure analysis considerations were covered by redundancy in the mounting of the front of the box section with 6 bolt positions, any 4 of which would have been sufficient, and low stress levels in the remainder of the box section and cantilever arms, the whole being stronger and stiffer than the original swinging arm mounting. The frame extension gave no problems during running.

This bike had all the characteristics of most attempts at very low seat feet forward machines in that it was difficult to ride very slowly, due, I think, partly to the inability to move rider weight about in the same way as we do unconsciously on a normal bike but mainly because of the lack of confidence caused by the knowledge that, if the bike leaned more than was easily corrected at low speed, simply putting a foot down on the ground for support, as you would do from a normal riding position, is not possible because your leg is at the wrong angle to provide the necessary support. In other words, they are the exact opposite of a trials bike where a solid dab can recover an almost impossible situation.

It was at its best on a fast main road run with little traffic and one good run at unmentionable speeds from the port of Heysham to near Glasgow at night returning from the IOM TT week sticks in my mind. It needed much less throttle to maintain a speed than the normal RD400 due to much reduced frontal area and the comfortable riding position and reduced wind blast encouraged maintaining a good speed for long distances.

Just one final thought on the general layout of the bike and all parts attached to the frame is the general need for lean angle clearance when cornering, taking into account suspension movement which is compressed by G-forces during cornering and adverse road cambers. I always try to have at least 45° lean to any hard contact with the bike static and unloaded. This may seem like a racing lean angle but, when you take off the effect of suspension compression and possible adverse road camber, raised manhole covers etc., it can easily be only 35° of usable clearance. Photo 29 shows what approximately 30° of lean looks like and shows about 2 inches of footrest clearance on a smooth surface.

Another point is that the position where final grounding occurs should be considered. If the first solid contact which could support the bike's weight is in front of the centre of gravity, the front wheel will tend to be lifted, leaving no possibility of regaining control and the bike going straight on until you hit something. It is much better to ensure that the first point of solid grounding is behind the centre of gravity so that the front wheel stays in contact with the ground, giving some hope of regaining control. On this subject I think the practice of having low forward mounted solid footrests together with wide tyres, which increase the lean angle at the same cornering speed, as found on some cruisers, is particularly bad unless some similar substantial low mounted parts are also fitted at the rear. You could call it "an accident waiting to happen" if any mildly enthusiastic cornering is attempted. (If you watch the policemen falling off in any old American car chase movie, the effect of grounding a forward footrest is obvious with the bike going into a corner, grounding a forward footrest, lifting the front and continuing to slid straight on, on its side.)

Photo 29 – 30° Lean on a smooth surface with fairly hard suspension.

*My feet forward had forward footrests that did not quite meet the 45° criteria although they were much higher than on many cruiser type bikes. I covered this potential **Hazard**, after much thought, by adding two lightweight go-kart wheels under the seat, set at an angle so that they touched down, when the bike leaned, just before the footrests. The theory was that they transformed the bike into an "instant sidecar outfit" which could continue to roll forward on 3 wheels and be recovered to vertical by a bit of reversed counter steer. I was never brave enough* *to seriously test the theory at high speed, other than a few light touches which it rode over with no problems. I did a test at low speed, starting with the bike leaned over onto the 3rd wheel. It proved easy to bring the bike upright when moving at quite a slow speed so I believe it would have been even easier when travelling faster. Photos 17 and 18 show the arrangement.*

This is a rather extreme example of how a design change can eliminate a ***Hazard*** *when a sufficiently low probability of it happening cannot be justified.*

An incidental advantage of this arrangement was that it gave the rider a lot more confidence when riding very slowly. Knowing that the possibility of uncontrollable excess lean is covered by the 3rd wheel, and you are not going to end up on one side, possibly trapped under the bike makes for more confident slow riding which, in itself, makes the possibility of losing control much less likely.

Still off the subject, but related to lean, measuring a large number of press photographs of Moto GP and Moto 2 riders at maximum lean, shows a remarkable consistency at around 55° to 56° from the vertical, measured from the tyre contact point through the estimate centre of gravity of bike plus rider, illustrating the much greater grip provided by hot racing rubber compounds. The angle measured along the bike centre line is a few degrees less, but it must make the design of footrests etc., much more difficult, needing to provide this level of clearance without an impossibly cramped riding position. There are exceptions with a small number of top riders who, in some photographs occasionally get to much higher (impossible), levels of lean, as if they operate on different laws of physics from everyone else. Perhaps they go well past the normal limits of adhesion and are much better at recovering from situations where most riders would inevitably crash.

Back to the subject yet again.

The following check list is included as a guide to the frame items which require some thought to avoid possible **Hazards**. Anything which is not directly taken from an equivalent production machine should be reviewed to see if there are any potential **Hazards** associated with it and to justify a low probability of any **Hazardous** failure.

Frame		
Type of failure & possible effects	**Possible causes**	**Typical justifications for a satisfactory low probability of failure**
Structural fatigue failure Possible effects: Loss of control of steering or brakes or inability to balance	Insufficient fatigue life due to: High stress cyclic levels in normal use Poor quality material Poor weld quality Corrosion Re-use following overload, e.g. accident damage.	Why will the frame not be prone to fatigue failure? Comparison with known satisfactory production frame Justification of material quality and inspection for any existing cracking Inspection for corrosion or damage Protection against future corrosion

Failure due to overload Possible effects: Loss of control of steering or brakes or inability to balance	Impact Loads e.g. wheelies, jumps, riding up high curbs, etc. These should not be a problem with a well-designed frame as a single event, but frequent repeated exceptional loads could result in undetected fatigue damage and eventual failure.	What would fail first in these cases and what would be the effect? It is often useful to rate the various features, in the load paths, in the order in which they would fail in the various overload cases. For example, we would not like the steering head to become detached from the frame as the first thing to fail with a vertical overload from the front wheel (heavy landing). A safer design would be for the suspension mounts to be less strong than the steering head attachment so that they would fail first and to provide an independent bump stop to prevent total suspension collapse locking the wheel.
Engine mount failure Possible effects: Engine mount failure is unlikely to be Hazardous providing that possible engine movement could not contact the road, open the throttle by pulling on the control cable or lock the back wheel	As above	As above It is usually quite easy to arrange a means by which any single failure of one of the engine mount positions is covered by redistribution of the loads through the other mounts or by some other means of support, thus making the failure non-hazardous.
Overheating Possible effects: Unlikely to be a problem with metal frames. Composite frames may lose integrity if severely overheated resulting in frame breakage and possible in loss of control of steering or brakes or inability to balance.	Overheating sufficient to affect strength is unlikely with a steel frame but strength of aluminium alloys and composites is significantly reduced. Overheating to temperatures above about 200°C for aluminium or 150° for most composites, which could result from close proximity to a motorcycle exhaust system (the exhaust gas temperature is much higher than this at high power but the metal temperature of the pipes is reduced somewhat by the cooling provided by external air flow). The exhaust is the only external part of a motorcycle where temperatures are high enough to affect the strength of frame materials in normal running, excluding the possibility of fire.	Are there any aluminium or composite frame parts close to or touching hot exhaust pipes at positions where there is significant stress in the frame? If so, how hot could they get in the worst conditions and, if marginally unacceptable temperature is possible, could they be insulated from the hot surface or moved, or the exhaust pipe run changed to provide clearance? If there are any exhaust supports from the frame, the possibility of heat conduction through the support should be considered, particularly at points close to the engine exhaust ports.

Lean angle clearance Possible effects: Hard contact with the road during cornering could reduce weight on the front or rear wheel resulting in loss of adhesion and resulting in loss of control.	Solid contact with the road when at maximum lean.	Are there any parts with insufficient lean angle clearance from the road, taking into account the effects of suspension compression and adverse road cambers, etc.? If so, could clearance be increased or the low parts, e.g. footrests, be modified so that they will hinge up if they contact the road? If the first contact is in front of the centre of gravity of the bike, could the design be changed to eliminate potentially Hazardous front wheel lifting if there was solid ground contact during cornering?
Excessive vibration Possible effects: Fatigue damage which could result in fatigue failure as above	Vibration can result in high cyclic stress levels in parts of the frame. A large number of load cycles at relatively low stress levels could cause a fatigue failure. The only likely cause of sufficient sustained vibration likely to cause fatigue damage is the engine, so engine mounting positions are the most likely to be affected. On classic singles or 360° vertical twins without a balance shaft, it may be possible to reduce vibration by a change of balance factor and/or dynamic balancing of the rotating assembly.	Are there high levels of vibration felt through the frame when the engine is running? If so, would it be possible to incorporate some rubber into the engine mount positions? If not, ensure that the engine mounts have sufficient fatigue strength, particularly if failure would be **Hazardous**. It may be possible to change the mounting system, such as adding a cylinder head steady link, which may alter the vibrating pattern and could also be arranged to ensure that failure at one mount position would not be **Hazardous**. Increasing the balance factor directs a larger proportion of the vibration in the direction at 90° to the piston stroke. Dynamic balancing equalises the vibration transmitted through both ends of the crankshaft eliminating any rocking component of the overall out of balance forces.
Corrosion Possible effects: Loss of control of steering or brakes or inability to balance	Corrosion at **Critical** points on the frame can result in fatigue cracking and eventual failure.	Are there any old, restored steel parts in the frame which may have undetected corrosion, particularly inside tubes? Are there any old magnesium parts in the frame which may have deep penetration corrosion not easily visible on the surface? If any corrodible parts are used at **Critical** positions, particularly in areas which cannot be seen on the assembled bike, are they sufficiently protected against corrosion? If severe corrosion is found, review remaining strength or look for another part.

4 – Rear Suspension

The best design of rear suspension seems to have been settled as some form of swinging arm pivoted at the front on the frame or gearbox casing and supporting the wheel spindle at the rear. All sorts of alternatives were tried in the 1950s including vertical plungers, short pivoted links, and the Triumph sprung hub which enclosed both spring and wheel location in a large wheel hub mounted in a rigid frame, but all had disadvantages compared the swinging arm. There were at least two early attempts at cantilever systems, the Vincent system which used a swinging arm with a near horizontal suspension units above the gearbox area, and the Douglas which used torsion bars inside the bottom frame rails, but most manufacturers eventually settled for swinging arms, initially with a spring shock absorber on each side close to the wheel centreline and, at a later date, adopted a single shock cantilever design with a single spring/damper unit mounted in front of the wheel with either a double or a single sided swinging arm.

In my opinion, there would seem to be no better alternative to using some form of swinging arm rear suspension on any modified or one-off motorcycle as it combines low un-sprung weight, sufficient stiffness and strength and a neat appearance, but special building is all about using ingenuity and trying something different, so who knows — alternatives might emerge.

There are basically 3 types of swinging arm suspension: cantilever, twin shock and single sided cantilever. See Sketch 14 below.

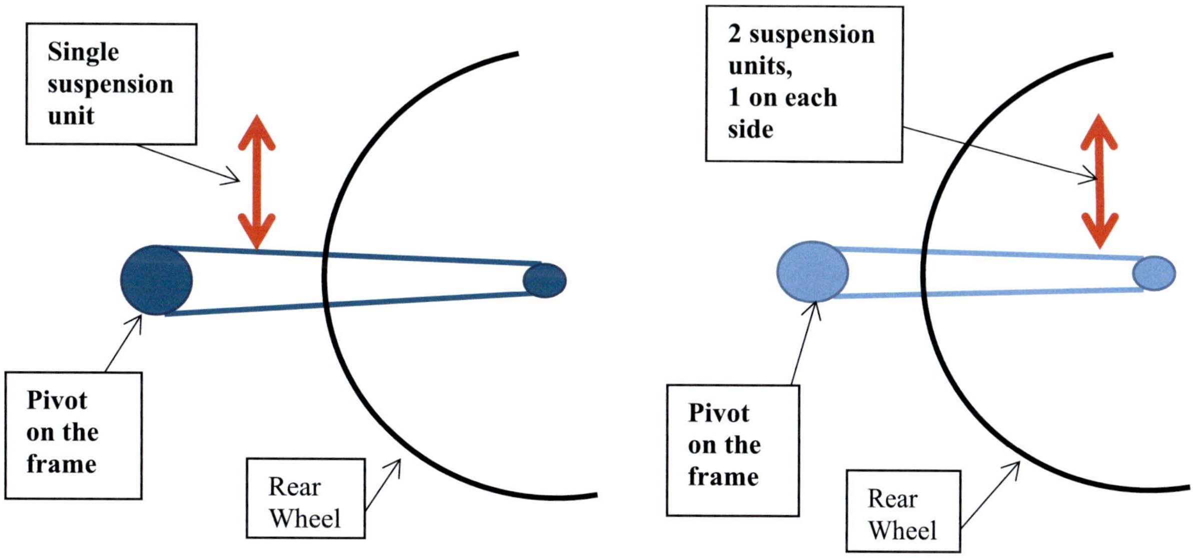

Cantilever swinging arm (single or double sided)
The spring load is applied to the swinging arm in front of the wheel

Twin shock swinging arm (almost always double sided)
The spring load is applied close to the wheel centre

Sketch14 – Types of swinging arm rear suspension.

Most modern bikes use a cantilever design with the spring shock absorber unit mounted in front of the wheel close to the swinging arm pivot on the frame. Direct mounting of the spring shock absorber unit at this position means very small spring movement for the required range of wheel deflection necessitating high spring rates (that is spring load increase per inch or centimetre of additional compression).

Various systems of link and levers have been designed to connect the swinging arm and the bottom of the suspension unit, either to increase spring movement and reduce the necessary spring rate, or to introduce rising rate where the effective spring rate at the wheel increases with increased wheel deflection. Rising rate has the advantage of providing a soft comfortable suspension on a relatively smooth road while reducing the chance of the suspension bottoming on a rougher surface but it tends to introduce more bearings at positions where they are exposed to a lot of road debris and water.

One way of achieving a rising effective spring rate at the wheel is to introduce a short lever into the system, linked to the swinging arm so that it to moves through a greater angle for a given swinging arm movement. The length of the short lever can then be used to vary the suspension unit movement in compression for any movement of the swinging arm and the greater angular deflection of the short lever can give the same effect of a rising rate spring. Sketch 15 shows the general principle although much neater designs are possible.

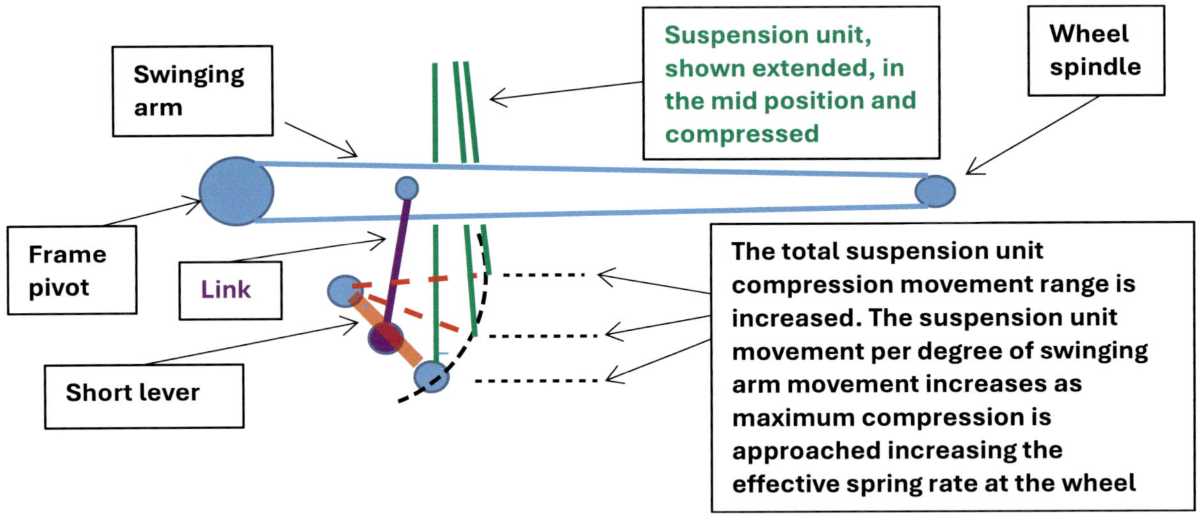

Sketch 15 – Principle of a rising rate link and lever system.

The other way of achieving a rising spring rate which is easier to incorporate in an existing suspension is to use a coil spring with the turns closer together at one end. As the turns close up during compression, the spring rate increases in proportion to the reduced number of "working" coils. Sketch 16 shows the general idea.

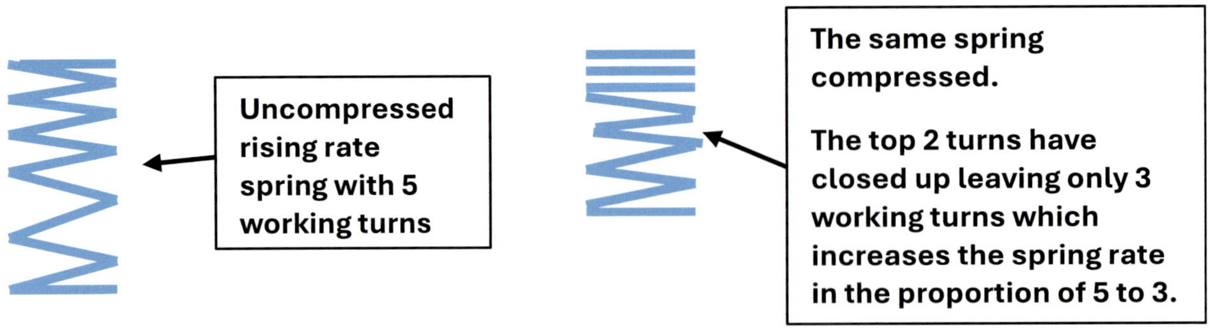

Sketch 16 – The principle of the rising rate spring.

A similar effect can be achieved with a conical coil spring as the larger diameter turns have the lowest spring rate and therefore close up first, but a conical spring is, in general, not as easy to fit into a motorcycle rear suspension, although often used on cars.

Cantilever suspension has many advantages in that it reduces un-sprung weight, helps weight centralisation and is visually a very clean neat design. The disadvantages are high local structural loads on the spring unit top and bottom mountings and the swinging arm pivot bearings, around 4 or 5 times the weight on the back wheel, and also the reliance on the single suspension unit and its mountings to prevent total suspension collapse.

Twin shock systems do have some advantages, the main one, from a failure analysis point of view, being that if one side fails and can no longer support the load, the other side unit will prevent total collapse. The other main advantage is that the weight of the rider, (plus pillion, plus luggage, etc,) is transmitted directly to the wheel through the springs, dramatically reducing loads on the swinging arm and rear parts of the frame. This could allow some weight saving in the swinging arm and frame compensating for the weight of the second spring unit.

Sketch 17 below shows the very high local loads on the suspension unit mountings and frame which result from the use of cantilever type suspension. These loads are transmitted through the swinging arm, necessitating a much stronger and stiffer swinging arm that is needed with a twin shock system.

The other loads transmitted through the swinging arm result from driving and braking forces and the bending and twisting effect of any side loads on the wheel at the tyre contact point.

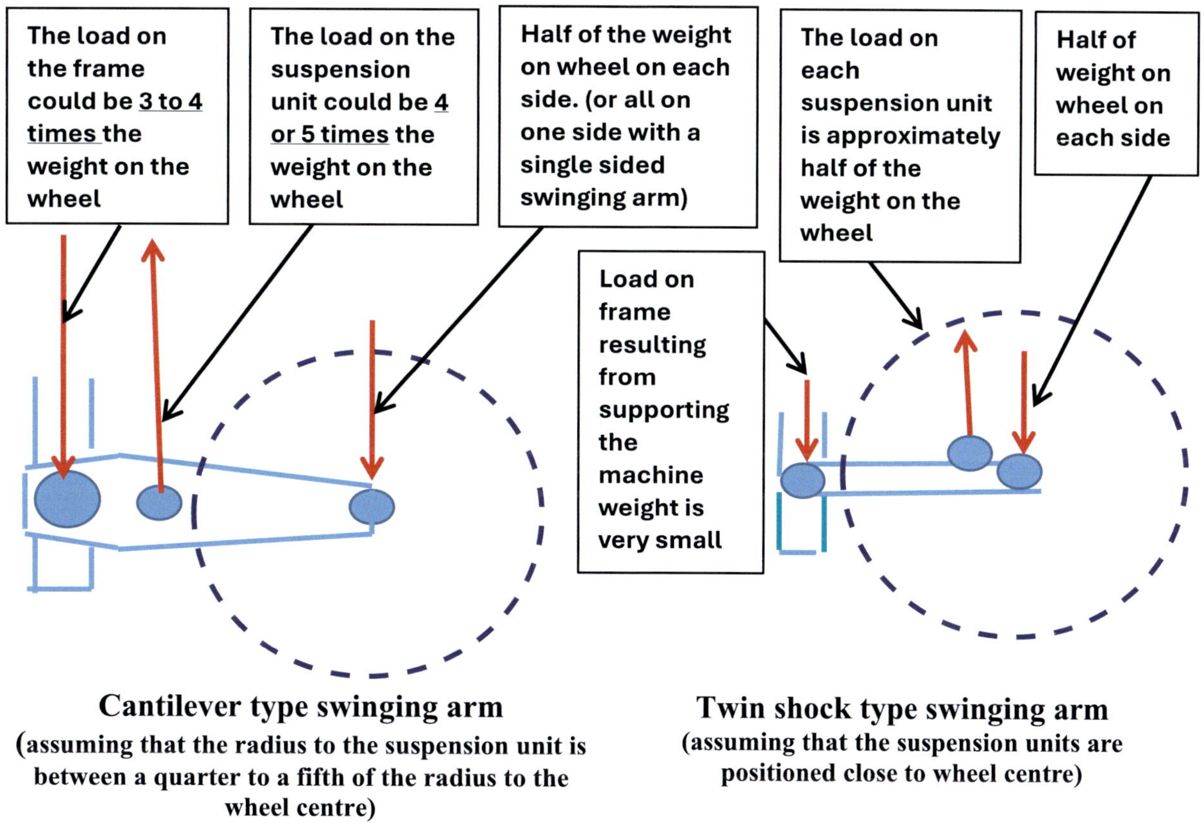

Sketch 17 – Structural loads associated with different types of swinging arm rear suspension.

Another design point on chain driven machines is that with a swinging arm, the distance between the gearbox and rear wheel sprockets varies as the suspension moves, making the chain slacker at the ends of suspension movement and tighter in the mid compression position. This effect can be minimised by positioning the swinging arm pivot as close as possible to the gearbox sprocket and on the straight line between the gearbox sprocket centre line and the rear wheel centre with the wheel at the mid position of the suspension travel. Sometimes, on high-powered machines, the height of the swinging arm pivot is varied to adjust the effect of chain loads on rear suspension compression or extension, to the detriment of constant chain tension.

A high pivot point reduces the effect of the high tension in the top run of the chain during hard acceleration which acts to further compress the suspension unit when weight transfer will also be acting to compress the rear suspension.

The parallelogram design which I used on the Horizontal Triumph (Photos 14 and 15) is no different in its operation from a normal swinging arm and was adopted for the practical reason of avoiding an excessively long wheelbase. The forward-facing cylinders increased the length of the engine/gearbox unit leaving no room for a swinging arm pivot behind the gearbox. It results in the disadvantage of increased un-sprung weight but does have the small advantage of reducing the variations in chain tension which result from the suspension movement of a conventional swinging arm because the effective centre of movement of the wheel spindle is close to the gearbox sprocket centre line.

Back to the subject of failure analysis. As with all other parts, we should look at any modified or unique parts in our suspension to see where they could fail, what would be the effect of such a failure, and, if it could be **Hazardous**, how do we justify an acceptably low probability of failure?

One point which may not be immediately obvious is the importance of the integrity of the chain adjusters. You would think that the only requirement for chain adjusters is to provide a means of setting the chain free movement before tightening the spindle nuts but if you consider the effects of a loosening spindle nut, the only thing then preventing the wheel being pulled forward by chain tension at the chain side is the chain adjuster. The wheel could then be pulled out of line by either high chain tension or heavy braking and possibly locked by contact between the tyre and the swinging arm. That is to say that a substantial chain adjuster will avoid a potential **Hazard** in the, not too unlikely, event of a loose wheel spindle nut. Eccentric type adjusters or those which push the wheel rearwards using a screw thread in front of the spindle are generally stronger that the type used on most classic bikes which use the screw thread to pull the wheel rearwards from the end of the swinging arm.

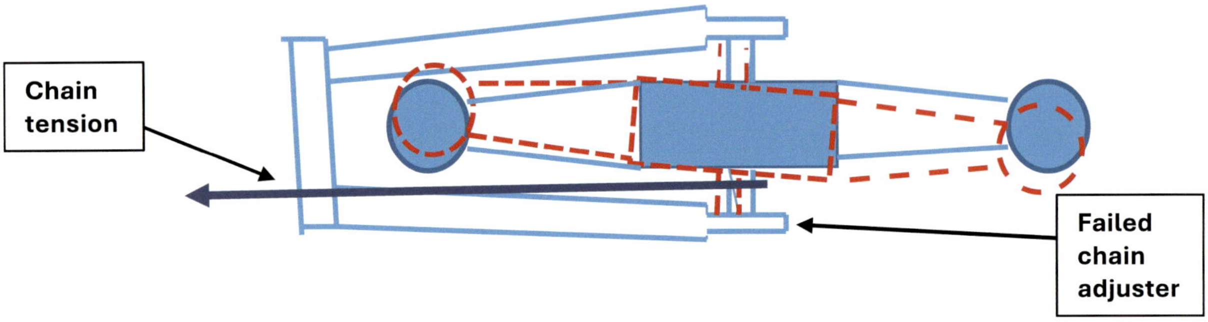

Sketch 18 – The effect of a failed chain adjuster allowing chain tension to affect wheel alignment.

If we review the failure analysis of a rear suspension system, the main potentially **Hazardous** failure types are likely to be those which may result in loss of correct location of the wheel or those which result in suspension collapse or wheel locking. These 3 failure categories are applicable to all types of suspension system but the means of justification of safety will depend on the individual design.

The following check list is included as a guide to the items which require some thought to avoid possible **Hazards**. Anything which is not directly taken from an equivalent production machine should be considered.

Rear Suspension		
Type of failure & possible effects	**Possible causes**	**Typical justifications for a satisfactory low probability of failure**
Structural fatigue Possible effects: Loss of control possibly **Hazardous**	Insufficient fatigue life due to: • High stress levels in normal use. • Poor quality material • Poor weld quality • Re-use following a high overload, e.g. accident damage. • Fatigue failure developing following undetected overload damage, such as the repeated impact loads which could occur during jumps, stunts or riding up high curbs, etc.	These are all related to the strength of the parts relative to the loads applied and the quality of the material and manufacturing processes. Have you reviewed the design to determine where a failure would make the bike uncontrollable and would therefore be **Hazardous**? If not, suggest a few minutes' thought would be worthwhile. Is there a good justification for a low probability of failure at all of the Critical positions identified above? If not, improve the design as necessary.
Suspension unit or unit mount failure Effect depends on whether free rotation of the wheel is possible after failure. If not, loss of control likely and would be **Hazardous**.	Fatigue at mount brackets or loss of mounting bolts.	Is there a good justification for the strength of suspension mounting brackets, and secure locking of mounting bolts, etc.? If not, improve the design as necessary. What prevents total collapse of the suspension until the tyre contacts some static part and locks the wheel, for example, if a suspension mounting bracket fails or a bolt is lost? If nothing would prevent total collapse suggest adding some form of bump stop of last resort. Are the suspension units of a standard production type in use on other machines? If not, are they of equivalent strength to typical production units?

Insufficient chain adjuster strength. No effect unless spindle nut becomes loose. If it does, possible wheel locking resulting in loss of control which would be **Hazardous**.	If wheel spindle nut loosens, failure of chain adjuster may allow the wheel to be pulled out of line by chain tension and possibly locked by contact with swinging arm or other static parts.	Are the chain adjusters strong enough to withstand the maximum chain tension? If not, suggest improving design. *As a guide, chain tension cannot exceed that required to break traction apart from momentary shock loads. Assuming a road tyre is used, the maximum possible traction force at the tyre is unlikely to exceed the weight on the wheel, including any weight transfer due to acceleration, so if the total weight of bike and rider is assumed to be the maximum weight on the wheel there will be a margin of safety.* *Then maximum possible chain tension will be approximately equal to the weight on the wheel multiplied by the ratio of tyre diameter to chain sprocket diameter, normally somewhere between 2 to 4. (a big number with a heavy bike capable of pulling a wheelie).* *(Racing tyre compounds will increase this but, I would guess, not by more than about 50%. Ref section 15.)* Are the chain adjusters securely locked after setting so that they cannot become loose, leaving total reliance on the tightness of the spindle nuts to maintain wheel alignment? If not, provide a suitable locking mechanism. A second nut acting as a lock nut is the normal method of locking a screw threaded type adjuster, although a self-locking nut would be generally satisfactory.

5 – Engine

Aircraft engine certification rules define the main types of **Hazardous** failure as falling into seven categories which can be summarized as follows:

1) Non-containment of high energy debris

   Failed parts may be ejected from the engine and damage other parts of the aircraft, for instance, a failed turbine or compressor disc could have enough energy at high engine speed to go through the engine casing, the engine cowling and the aircraft fuselage and come out of the other side.

2) Concentration of toxic products in engine bleed air used for cabin pressurisation

   It is not good to poison the passengers and crew with toxic fumes in the cabin air which is taken from the engine compressors in most aircraft. The most likely source of toxic fumes is engine oil leaking into the very hot compressor bleed air.

3) Significant thrust in the opposite direction to that commanded by pilot

   This is mainly aimed at the possibility of an engine going into reverse thrust in flight, which would generate high levels of drag and could make it impossible to control the aircraft or maintain height.

4) Uncontrolled fire

   Most possible engine fire situations would be detected and there would be procedures in place to control and extinguish the fire, but a fire in flight which could not be extinguished would be an obvious **Hazard**.

5) Failure of the engine mount system

   If the engine mount system failed, the engine could fall off the aircraft, an obvious **Hazard** both to safe flight and to anyone in the "wrong place" on the ground.

6) Release of propeller

   Release of a propeller is similar to **Hazards** 1) and 5) in that a large, released spinning disc could go anywhere and could do a lot of damage if it hit any part of the aircraft or fell to the ground below.

7) Inability to shut engine down

   An engine which cannot be shut down would be an obvious **Hazard** when attempting to reduce speed or altitude or to land.

Most other types of engine failure would normally fall into either **Major** or **Minor** categories on a multi-engined aircraft as they would not affect the remainder of the aircraft, which could continue to fly on the other engine or engines even if the failed engine had completely destroyed itself. Single engine aircraft are in a different category where maintaining stability and controllability in gliding flight is the least severe consequence of engine failure.

Of these **Hazardous** aircraft engine failure categories, the only ones possibly applicable to a motorcycle are numbers 1, 4, 5 and 7.

Number 1 relates to engine parts which, if they fail, could come out through the engine casing, such as a failed connecting rod but with a conventional motorcycle layout, debris from these types of failure would not generally be directed towards the rider and so would not be **Hazardous**. Racing sidecars where the rider is directly over the engine with only a thin engine cover separating them would be a bit more exposed to the possibility of a major engine blow up ejecting things upwards.

About the only parts of a normal motorcycle design where there is potential for injury are flywheel generators of the rotating magnet type which have heavy rims and are usually positioned on the end of the crankshaft and perhaps aimed at the rider's leg. I have never heard of one failing and firing bits of flywheel rim in all directions, but this is probably due to large strength reserves and good manufacturing quality on the flywheel generators of production machines. Lightening the rim of this type of generator to reduce flywheel inertia and improve engine acceleration is definitely not recommended.

An interesting thought on this subject is that almost all production front wheel drive cars have large heavy flywheels storing nearly as much energy as many aero engine turbine discs and are aimed directly at the driver or passenger's legs. Again, I have never heard of one failing, but it is still a potential ***Hazard*** *even though the probability of it happening appears to be very low.*

From the motorcycle engine modification point of view, the most obvious point is that no flywheel type component with any sort of damage should be used and any modifications to a flywheel should be carefully considered so that stress levels and not increased. The most usual modifications would be for balancing or to reduce weight, reducing inertia to increase acceleration at the expense of more low speed roughness.

On the subject of balancing, the practice of drilling holes in the flywheel rim, as was common on classic machines is, in my opinion, a very bad idea as it introduces stress concentration in a fairly random manner. It is much better to remove metal from the surface of the outer diameter or the sides of the disc rim without introducing any new stress concentrating features such as drilled holes or sharp edges. Alternatively, it may be possible to add balance weights to any existing bolt holes but new bolt holes for this purpose are not good for the same reason as drilled balance holes.

Flywheel weight reduction should not be attempted without some knowledge of the stress distribution in the flywheel. With a plane parallel sided disc, removal of some metal from the outer diameter or sides of the disc rim will be reasonably safe but changes at the inner diameter should be avoided.

The greatest power gains resulting from engine modifications for competition purposes will always come from increases in maximum RPM. If you think about it, a 200 horsepower 1000 cc superbike produces its power at roughly twice the engine speed of a 50 horsepower classic 500 cc single such as the Manx Norton and is twice the capacity, meaning that the power generated per litre per engine revolution is about the same.

From the failure analysis point of view, it is worth noting that the stress levels in a rotating disc will increase with the square of the speed. That is a 10% speed increase say from 10,000 to 11,000 rpm would result in a 21% increase in stress levels in all the main rotating and reciprocating parts, all other things being equal, with a consequent increase in the risk of failure.

Off the subject again, I once saw a lightened flywheel for a racing BL Mini offered for sale which had large elongated machined holes through the disc between the clutch face and the hub, making the flywheel more like a spoked wheel. The average motorcycle road wheel runs at just over 800 rpm at 100 mph and a racing Mini engine could run up to 10 times that speed.

Centrifugal loads are only a small part of the loads on a road wheel, as these loads mainly result from the weight of the bike, but as centrifugal stress levels rise with the square of the speed of rotation, the centrifugal stress levels in the same wheel would be 100 times higher at 8,000 rpm than at 800 rpm. The lightened flywheel would be much stronger than a road wheel, but I hope whoever designed this one did some detailed stress calculations as, on the old principle of "What looks right generally is right", (which is usually but not always true), it definitely looked wrong, and the old Mini flywheel is positioned directly in line with the driver. The flywheel rim speed could be in excess of 300 ft/sec (204 mph), so the thin aluminium flywheel casing and the sheet steel bulkhead are unlikely to provide much protection against a large piece of flywheel ejected at this speed.

There is one aspect of release of material from inside the engine which is not generally **Hazardous** on a multi-engined aircraft, as it would, at worst, only cause the engine to be shut down, but could be **Hazardous** on a motorcycle. A sudden loss of large quantities of oil from the engine could go directly onto the rear tyre. In general, rapid loss of oil as a result of failure of the main engine casings is very unlikely but there are sometimes parts of the oil system such as pressure switches, oil pipe connections filters or oil coolers which are mounted in positions which are vulnerable to damage from flying debris or contact with the road surface or just general deterioration with age. A check for vulnerable parts is easy to do and anything suspect can generally be moved or provided with some sort of a guard eliminating a possible **Hazard**.

I have not included Number 3 in the aircraft engine rules; thrust in the opposite subject direction to that intended, but it could be argued that this is equivalent to motorcycle engine seizure locking the wheel and causing braking where constant speed or acceleration is intended. The difference is that the motorcycle clutch allows the locked wheel situation to be corrected whereas an engine stuck in reverse thrust on an aircraft could be beyond the pilot's control. Thus, barring the unlikely event of a sudden engine seizure when at a high lean angle, with no prior warning, engine seizure on a motorcycle is unlikely to be directly **Hazardous**.

No 4, an uncontrolled fire, is covered separately but any fire on a motorcycle would be uncontrolled as there is no on-board means of putting it out. The saving fact with respect to **Hazards** is that a motorcycle can be quickly stopped and abandoned whereas you are stuck with an uncontrollable aircraft fire until a landing can be achieved.

Number 5, failure of the engine mounting system, is an obvious **Hazard** on an aircraft with a heavy engine released in the air which may go anywhere. On most motorcycles there are several engine mount positions and failure of one, for example the bolt coming out, would not generally let the engine move very far out of position. The effects of mount failure may be more severe or even **Hazardous** for engines which form part of the frame or are suspended below the frame, but every design is different in this respect and the effect of disconnecting each mount in turn should be reviewed to make sure that something would catch the engine before it dropped to drag along the road.

Number 7, inability to shut the engine down, is also an obvious **Hazard** on an aircraft and is normally countered by provision of a fuel shut-off valve for each engine which is totally independent of all other engine controls. The equivalent on a motorcycle is a stuck open throttle which can be countered by a handlebar stop switch, provided that it is easily accessible without letting go of the handlebar and cuts the ignition directly with no reliance on any other part of the control system.

The following check list is included as a guide to the items which require some thought to avoid possible **Hazards**. Anything which is not directly taken from an equivalent production machine should be considered.

Engine		
Type of failure & possible effects	**Possible causes**	**Typical justifications for a satisfactory low probability of failure**
Uncontained failure of rotating or reciprocating components Effects: An engine failure should not be **Hazardous** unless some parts are ejected at high speed in a direction which could cause injury to the rider or anyone else.	Fatigue originating from modifications or pre-existing damage	Are there any modifications to internal engine parts relative to manufacturer's specifications? If so, ensure that the modified parts are equivalent or better than the standard parts in respect of strength and fatigue resistance at any position where failure could be **Hazardous**. Have all used parts fitted been inspected for cracking or damage? Fitting a damaged part is likely to increase the risk of failure. Is it proposed to increase the maximum engine speed relative to the production use of the engine, or to add a supercharger? This sort of modification increases stress levels in the main rotating and reciprocating engine parts and the effect on the probability of a failure of every Critical part should be considered.
Failure of engine mountings Effect: Engine may move out of position, possibly interfering with control which could be **Hazardous**.	Fatigue or over-load failure of one engine mounting, or a lost engine mounting bolt.	Where would the engine move to if any one of the engine mounts failed or any mounting bolt is lost? If the other engine mounts don't prevent movement to an unsafe position, e.g. moving down until it contacts the road, can some sort of secondary mount be provided to avoid a **Hazard**?
	Engine movement out of position affects throttle control. or overloads oil or fuel pipe connections	If any mount position becomes detached as above, does the movement of the engine put any load on the control cables or oil and fuel pipes, which may result in a **Hazard**? If so, can the **Hazard** be designed out by slightly longer cables or pipes or by preventing sufficient movement of the engine?
Oil leaks Possible contamination of rear tyre. Potentially **Hazardous**	Minor oil leaks will have no **Hazardous** effects but the most likely reasons for a substantial oil leak are leaking external oil pipes, and their connections, or loss of the sump drain plug	Are all external oil pipes in good condition with secure connections and suitable clipping support to prevent vibration without restricting any possible thermal expansion? If not, review system and improve as necessary. All pipes connecting the engine to other frame mounted equipment should have a flexible section. Is the drain plug unchanged from a standard production design with screw threads in good condition and correctly tightened? If not, repair or replace as necessary. Consider wire locking, particularly if the arrangement is non-standard.

6 – Clutch and Gearbox

On a motorcycle, the clutch is, in effect, another flywheel and is subject to the same considerations and, in some cases, is exposed close to the rider's legs with little protection provided by enclosure. Care therefore needs to be taken to ensure that any changes made to the clutch do not make cracking of the drum more likely as flying pieces of clutch drum would definitely be **Hazardous**. The only other obvious potentially **Hazardous** failure of the clutch would be loss of the retaining nut allowing the whole unit to come off the shaft. The method of locking used on this nut needs to be reviewed to ensure that it is secure.

There is nothing much to learn from aircraft engines on the subject of clutches, as the use of clutches is usually limited one-way clutches in the drive to or from some accessories such as the engine starter. In the same way as a motorcycle electric starter, the starter needs to drive the engine during starting, but the engine when running must not be able to drive the starter. Failure to disconnect the starter when the engine has started is potentially **Hazardous** as the gear ratio between the starter and the engine main shaft will be such that at maximum engine speed, the starter speed would be impossibly high with the potential for complete disintegration of the starter and possible **Hazardous** release of high energy debris.

This type of **Hazard** would normally be designed out by some means such as reducing the strength of the starter shaft so that it would shear off before the starter failed, or by some similar means.

The opposite potential **Hazard** could be if the starter, which needs to be extremely powerful on a large gas turbine, fails to connect to the engine during an attempted start and runs free. In this case the starter, if it is the normal compressed air operated turbine unit, will immediately run up to an extremely high speed and Engine Certification requires that no **Hazardous** failure of the starter rotor should result from running free.

Some modern aircraft starters are designed to also operate as electrical generators to combine the two units and save weight, but some clever design is needed to ensure that the starter/generator runs at safe speeds in both generating and starting modes of operation.

Failure to disengage the starter after the engine has started on a motorcycle would have a similar effect of over-speeding the starter but in general the starter parts would be relatively small and light, and a failed starter or starter drive may be expensive but is unlikely to be a **Hazard**.

The gearbox on a motorcycle, unlike the engine, which is protected by the clutch, is directly connected to the rear wheel so seizure of the gearbox would lock the rear wheel and therefore be **Hazardous**. Motorcycle gear trains and their bearings are generally very robust, and failure of a production box is unlikely even when subject to increased power.

The gearbox can only transmit sufficient torque to either spin the rear wheel or lift the front wheel, so gearbox loads are self-limiting regardless of how powerful the engine is.

There are basically two types of gearbox failure which could be **Hazardous**. A loose bit of a broken gear tooth or engagement dog or other loose object which could get caught between another pair of gears locking the gear train, or a failure of the selection mechanism allowing two gears to be selected at the same time.

Checks during assembly and fitting to ensure that no loose debris is allowed to remain in the box are therefore essential. It is also worth checking selector forks for potential cracking, distortion, or wear. Any sharp edges, etc, which could result in stress concentrations and thus possible fatigue origins, should be radiused, but no other metal removal should be attempted.

A slipper clutch is designed to prevent rear wheel locking as a result of engine braking and would help in a case of total engine seizure, but unfortunately provides no protection for the gear box.

One possibility would be to separate the slipper mechanism from the clutch and fit a different mechanism, which achieved the same effect, in the gearbox output shaft or even in the rear wheel hub. Slip could then be matched more closely to the available tyre to road adhesion, independent of the gear selected, as well as providing protection in the unlikely case of gear box lock up.

One possible mechanism would be to have a one-way clutch, directed to allow free forward running, driving through a single plate normal type of clutch arranged to lock the one-way clutch, thus preventing free running and allowed full engine braking in normal operation. This clutch could then be partially lifted to allow some slip when required, either mechanically, in a similar manner to existing slipper clutches, or perhaps electrically which would allow programming of the operation to suit speed and road conditions.

To the best of my knowledge, no one has tried this idea to date but it could provide a real step forward on a high-powered bike as the amount of slip could be directly related to chain tension and thus to tyre to road adhesion.

It seems strange to me that very few motorcycle gearboxes are equipped with a magnetic drain plug which would hold any loose debris and prevent it getting into the gear train. A loose piece from a chipped gear tooth retained by a magnetic plug would be totally safe but floating around in the gear oil, it could be a potential **Hazard**. Engines are usually fitted with this type of magnetic debris collector and an oil filter, although from a failure analysis point of view, engine seizure is a less severe failure than gearbox seizure. If the gearbox is not already fitted with a magnetic plug, a good safety modification would be to fit one. It should be an easy modification to do and may just avoid a potential **Hazard** gearbox seizure at some time. In any case it is good to collect any metallic debris that may be floating around in the gearbox oil.

As always, thinking about failure analysis, a bit of caution is needed before fitting a magnet to a drain plug. If the magnet came out and was loose in the gearbox oil, it would be attracted to the steel gears and would be likely to become trapped between gear teeth, locking the gearbox and the back wheel, a safety system which could suddenly become a **Hazard** itself, not an ideal situation.

Attaching the magnet by fixing it with adhesive in a hole drilled in the drain plug, is therefore not a good idea. It needs some sort of more secure method of attachment to be safe.

One way would be to find an existing production magnetic plug, perhaps off a car. If no plug with the right thread could be found, it should be possible to get one with a slightly bigger diameter, machine it down and rethread it to suit.

If adhesive is the only practical way of attaching a magnet to the drain plug, I would use a steel plug so that the magnetic attraction helped retain the magnet, which would not be the case with a non-magnetic material such as brass. Use a cylindrical magnet in a blind hole with a close sliding fit (about 0,002 or 0.003 thou. [inches]) clearance in a steel plug. De-grease the magnet and the hole and put a few drops of something like "Locktite 648" retaining compound in the hole. Then press the magnet in so that the compound squeezes out through the clearance, ensuring a complete coating all around the magnet. I think you could then justify an acceptable probability of the magnet not coming adrift by this method.

The mention of clearances for adhesive bonding brings to mind another point, nothing to do with gearboxes, but worth mentioning. If two smooth flat surfaces are to be bonded together, clamping them tightly together while the adhesive cures can squeeze out most of the adhesive, leaving the joint much weaker than it should have been. Light, or no, clamping often risks the joint moving and bonding in the wrong position. One method I have seen used successfully is to cover one surface with an array of small centre pop marks which hold the surfaces a few thou apart allowing tight clamping without squeezing out the adhesive film and allowing the finished joint to have the full strength of the adhesive over all the area.

Back to the gearboxes.

Aircraft engines always have magnetic debris collectors in all the oil return pipes to the tank, which are inspected regularly for early signs of wear or failure of lubricated parts. On the more modern aircraft engines these magnetic detectors are often fitted with electronic sensors which continuously monitor the detector and flag up any abnormal debris collection automatically. Analysis of the material in the debris collected can often show where it has come from and identify which parts are wearing before the wear develops enough to affect running.

Back to the bikes. If there is a gearbox drain plug in the motorcycle gearbox, it is worth remembering that it is usually ideally placed to put oil onto the rear tyre if it loosened and was lost. Although there is no locking of the plug on most production machines, plug retention is generally satisfactory, but loss of oil at this position is potentially **Hazardous** and it is worth a thought as to how secure the plug is, particularly if some changes have been made in this area. Wire locking is easy to do and should be considered if any doubt exists.

Aircraft gearboxes are, in general, fixed ratio gear trains used to drive accessories such as pumps and generators or to reduce engine shaft speed to drive a propeller or large fan. The reason for reducing propeller speed is that the larger the propeller diameter gets, the faster is the blade speed at the tip for a given RPM. The air speed over the blade results from the combination of rotating tip speed and forward speed and an air speed much in excess of the speed of sound at the blade tip would be very inefficient.

Aircraft gearboxes are, in general, subject to the same rules as the engine and it would be unusual for there to be any **Critical parts** in an accessory gearbox. A propeller gearbox is subject to more potential risks due to the very high power transmitted and the large inertia of the propeller which would overcome any loose object trapped between the gear teeth, possibly springing the gears apart and breaking the gearbox casing. This may release the whole propeller or could applying a very large shock torque load to the engine mountings. The failure analysis of a propeller gearbox therefore needs very careful consideration as a detached propeller could go anywhere and cause a lot of damage, like a flying circular saw, and any failure of the engine mountings is potentially **Hazardous**.

I have seen a photograph of a large turbo-prop transport aircraft where one propeller had come off on a first ground test run after maintenance (it must not have been fitted correctly). It had hit the fuselage and cut through what seemed to be about a quarter of the circumference. If it had happened in flight, it is doubtful whether the aircraft would have survived.

There is very little totally new.

If we go back to the early days of the steam driven textile industry in Victorian times, there is an old Lancashire Mills tale of a mill engine flywheel which escaped from its bearings for some reason when the engine was running. It was said to have gone straight though the engine house wall and bowled down the street outside leaving a trail of devastation in its wake.

These Victorian mill engine houses were usually very solidly built, a surviving local one has solid stone walls almost 2 feet thick at the base, so it would take a very large impact to go through it. The steam boilers were usually not housed in the engine house, good failure analysis thinking, although they probably just called it plain common sense. If the escaping flywheel, or any other heavy "flying" engine part, had hit a pressurised boiler there would probably have been an enormous explosion which could have collapsed the whole building.

Whether the story is true or just a myth, these very heavy large diameter cast iron flywheels could easily store as much energy as an aircraft propeller, even though they ran at much lower speed.

I have strayed a long way from gearboxes, but these thoughts keep coming back.

Back to the present day and the gearboxes.

My only experience of a complete transmission lock up due to a gear problem was fortunately in a car, not on a bike, in the 1960s. The rear axle in a rear-wheel-drive car suddenly locked up at about 40 mph on a narrow country road. More by luck than management, it stayed between the hedges until it stopped and did not hit anything. The cause was a small gear shaft carrying the planet gears in the differential, which had moved out of position and hit the main pinion gear, taking a big piece of gear tooth out of the cast iron crown wheel. The gear shaft should have been retained in position by a locking pin, but a non-standard part had been fitted which did not fully lock the shaft in position. Another lesson learned, not to take any risks with gearbox parts.

Sketch 19 below is a good example of how the problem with a very small part, the differential gear shaft locking pin, can sometimes result in a ***Hazardous*** *failure.*

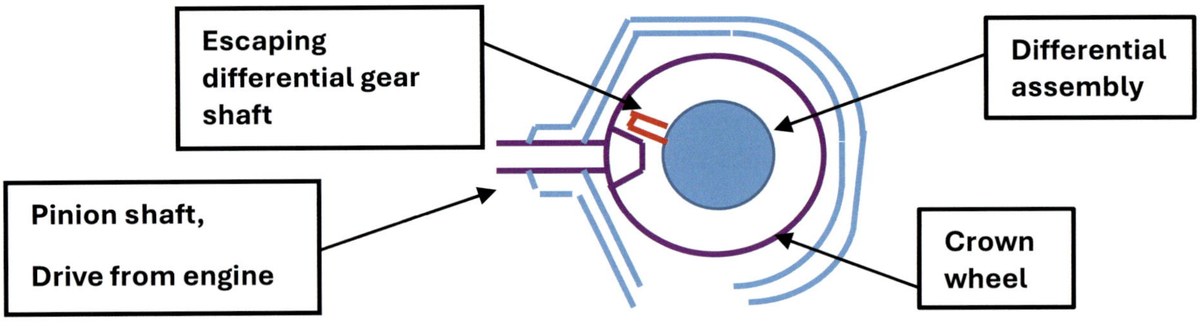

Sketch 19 – The rear axle gears showing the escaping differential gear shaft which locked the rear wheels.

The following table gives a guide to the sort of things which should be considered particularly if anything has been changed from a standard production arrangement.

Clutch and Gearbox		
Type of failure & possible effects	**Possible causes**	**Typical justifications for a satisfactory low probability of failure**
Clutch drum failure Pieces of drum released possibly penetrating casing, etc. and hitting rider. Possibly **Hazardous**	Fatigue cracking	Is the clutch drum a standard production item with no modifications apart from minor radiusing of sharp edges? If not, are you sure that the changes do not reduce strength?
Clutch retaining nut loose or lost. Whole clutch may be released, possibly causing injury. Possibly **Hazardous**	Screw thread in poor condition. Not tightened correctly. Nut locking system not correctly installed.	Has thread condition and tightness been confirmed? Is the nut locking system satisfactory? *(A deformable tab washer is probably the best locking system to use on this nut if there is enough access to install one.)*
Gearbox seizure Locked back wheel, **Hazardous**	Loose items between gear teeth. Selector failure allowing two gears to be selected at the same time.	Is the gearbox a standard production type with no internal modifications? If not, review changes for potential cracking and remove any stress raisers. Are you confident that the gearbox has not been abused in previous use with possible cracked gears or other parts which may release debris, possibly getting between other gear teeth and locking the drive? If not, strip the gearbox and inspect all parts.
Loss of gearbox drain plug Oil on rear tyre. Possible loss of control **Hazardous**	Screw thread in poor condition. Not tightened correctly.	Has thread condition and tightness been confirmed? Consider the desirability of wire locking, particularly if any aspect is non-standard. *A change to a magnetic plug if not fitted is a worthwhile simple modification which may avoid a seized gearbox one day, if any debris became detached inside the box and was circulating in the oil.*
Loss of magnet from a magnetic drain plug Magnet could become trapped between gears, locking the gearbox and rear wheel. **Hazardous**	Magnet not positively attached in the drain plug e.g. only fixed with adhesive.	Is the magnet positively located in the plug to ensure that it could never become detached? If not, don't fit it. Find a replacement or just fit a plain plug without a magnet.

7 – Chain and Belt Drives

The obvious failure to be considered on any chain or belt drive is what happens if the chain breaks, how it affects whatever is driven by the chain and where the loose chain goes.

Not many chain drives are used on aircraft, although they have been used to drive the swivelling exhaust nozzles, which alter the direction of the engine thrust between vertical for lift and horizontal for propulsion, on some vertical take-off aircraft, not the best place to have a chain failure but no doubt the chains used would have a lot of reserve strength and a thorough inspection and maintenance schedule.

On a motorcycle, the final drive chain is generally the only one where there is a possibility of **Hazardous** failure. Failure of any other chain drive, within the engine/gear box unit, is unlikely to have any **Hazardous** effects. The most likely effect of an internal chain failure, such as failure of a cam chain, will be total loss of power and stopping the engine. A primary drive chain failure would let the engine run free, perhaps up to the speed limiter, or to the valve float speed on an engine with no speed limiter, and the loose chain possibly may lock the drive to the clutch or possibly damage the chain case.

The exception to this is the exposed primary chain on some classic racers which could, in the worst case, fly off and hit the rider's leg. A strong chain guard to retain a broken primary chain would be a good idea on these bikes, well worth the small amount of extra weight on the day when it may prevent a serious injury.

Failure of a final drive chain will usually result in the chain running cleanly off the sprockets and onto the road, leaving the same conditions as would occur with the gear box in neutral, and therefore would not normally be a **Hazard.** I did hear of one case where the flying end of a broken final drive chain, with no chain guard fitted, hit the rider in the rear end causing quite serious injury. The failure analysis solution to the possibility of final drive chain failure is therefore to make sure that there is plenty of clearance for the chain to run off cleanly, particularly around the gearbox sprocket, and to provide a chain guard over the top run which is strong enough to contain the flying end of a broken chain.

The second most likely **Hazard** in the event of drive chain failure, would be that a broken chain could "ball up," (become folded such that the radial thickness attempting to pass round the sprocket was two or more times the radial thickness of the chain), and therefore could possibly become trapped and lock the back wheel. This is unlikely if the bottom run of the chain is open but, if some form of chain enclosure or a bottom chain guard were fitted, the broken chain could become trapped between the enclosure or chain guard and the bottom of the rear wheel sprocket, locking the rear wheel.

I included a bottom chain guard in my mildly modified 1954 Norton Dominator (Photo 16) and it proved to be a very good modification for improving chain life by keeping water and road debris off the chain. To cover the possibility of a broken chain balling up and locking the rear wheel, I made the bottom chain guard from very thin aluminium sheet which was not strong enough to contain a balled-up chain locking the wheel. From my experience with this modification, I concluded that the majority of road debris and water reaching the chain came from the front of the rear tyre onto the top the bottom run of the chain where the tyre is travelling downwards and passes the chain run. All my later machines have therefore just included an extension of the top chain guard between the tyre and the chain to a level below the bottom run. This small change, (just a piece of aluminium sheet riveted to the inner side of the chain guard), has proved to be almost as effective as the full bottom chain guard in keeping the chain in good condition. See Sketch 20.

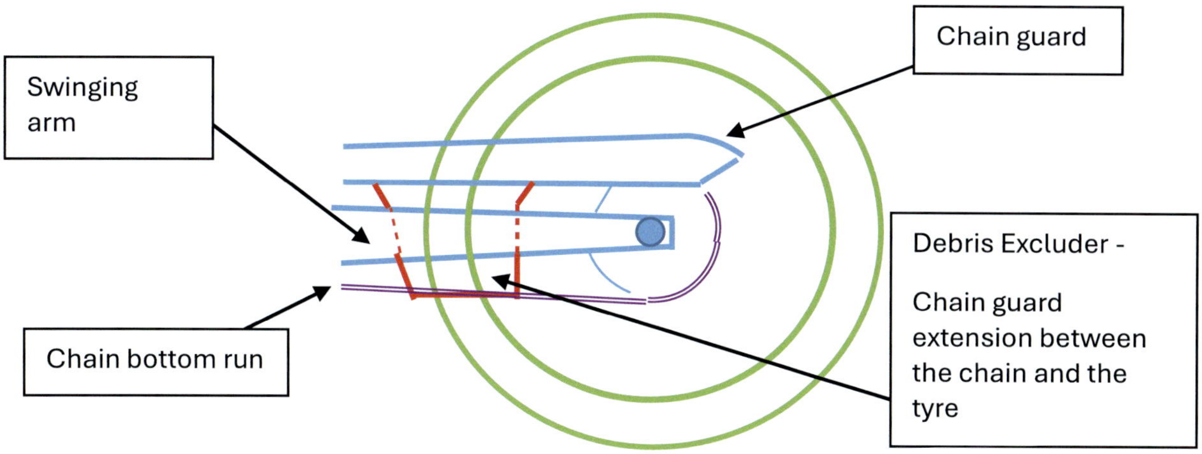

Sketch 20 – Debris excluder to prevent debris carried round by the tyre from falling onto the bottom run of the chain.

The only other possible **Hazard** which comes to mind, and could result from a broken final drive chain, is if the impact of the chain with the engine or gearbox casing or trapping of the chain between the sprocket and the casing resulted in holing of the casing, releasing a large quantity of oil onto the rear tyre. It is a good idea to check the clearance around the outside of the chain as it runs round the gearbox sprocket, particularly if a larger sprocket has been fitted to alter the gearing. Ideally, this clearance should be at least enough to allow a spare chain link to pass around the sprocket between the outside of the chain and the casing without touching the casing. If there is not enough clearance for this, I would consider a smaller gearbox sprocket and adjusting the gearing by means of a proportionally smaller rear wheel sprocket.

The types of failure suggested for chain drives also apply to belt drives with the exception that a broken belt is unlikely to lock the rear wheel by folding into a sufficiently large mass and would generally be of insufficient weight to seriously impact the engine casing. Therefore, from a failure analysis point of view, a belt drive has considerable advantages over a chain. From the design point of view, a belt drive has both advantages and disadvantages relative to a chain and they are becoming more common on production machines with improvements in belt strength and durability.

The main advantage of a belt drive is obviously the lubricant free nature of the drive which is cleaner and requires less maintenance than a chain. The disadvantage, where very high power is concerned, is that the necessary belt width becomes too large to be easily accommodated in the design, particularly in the final drive where belt tension is increased proportionally by the speed reduction in the gearbox. The maximum possible continuous belt tension is dictated by tyre adhesion rather than engine power but peaks above this can occur, due to the inertia of the rotating parts, with sudden acceleration or braking. A considerable margin above the maximum continuous belt tension should therefore be allowed when selecting a belt for a one-off belt drive system.

The system is definitely worth considering for a totally unique special, but conversion of an existing final drive chain to a belt drive is not a usually a practical proposition. The exception (there is always an exception to any general rule) is the primary drive from the engine to the gearbox on a classic bike where a belt conversion is definitely the best way to go and there are many kits on the market which fit most of the common classics.

Another thought on chain or belt drives on a classic bike with a separate engine and gearbox connected by a primary chain or belt drive is worth mentioning, although unlikely to cause a **Hazardous** failure. Chain adjustment on the primary drive is often set by means of a screwed adjuster which moves the gearbox forward or rearwards relative to the engine. Once the primary drive has been adjusted, the rear drive chain can be re-adjusted to take into account the new gearbox position. Tension in the rear chain is always much greater than in the primary chain because of the gear ratio between the engine and the wheel so that if the primary chain adjuster and the clamping of the gearbox in position is not adequate, the gearbox will be pulled rearwards. This would put a very high load on the primary chain which may eventually result in chain failure or bearing failure in the engine or gearbox. The integrity of the primary chain adjuster and the clamping of the gearbox to prevent movement is therefore an important feature on a bike with this type of drive. Sketch 21 below shows the usual arrangement.

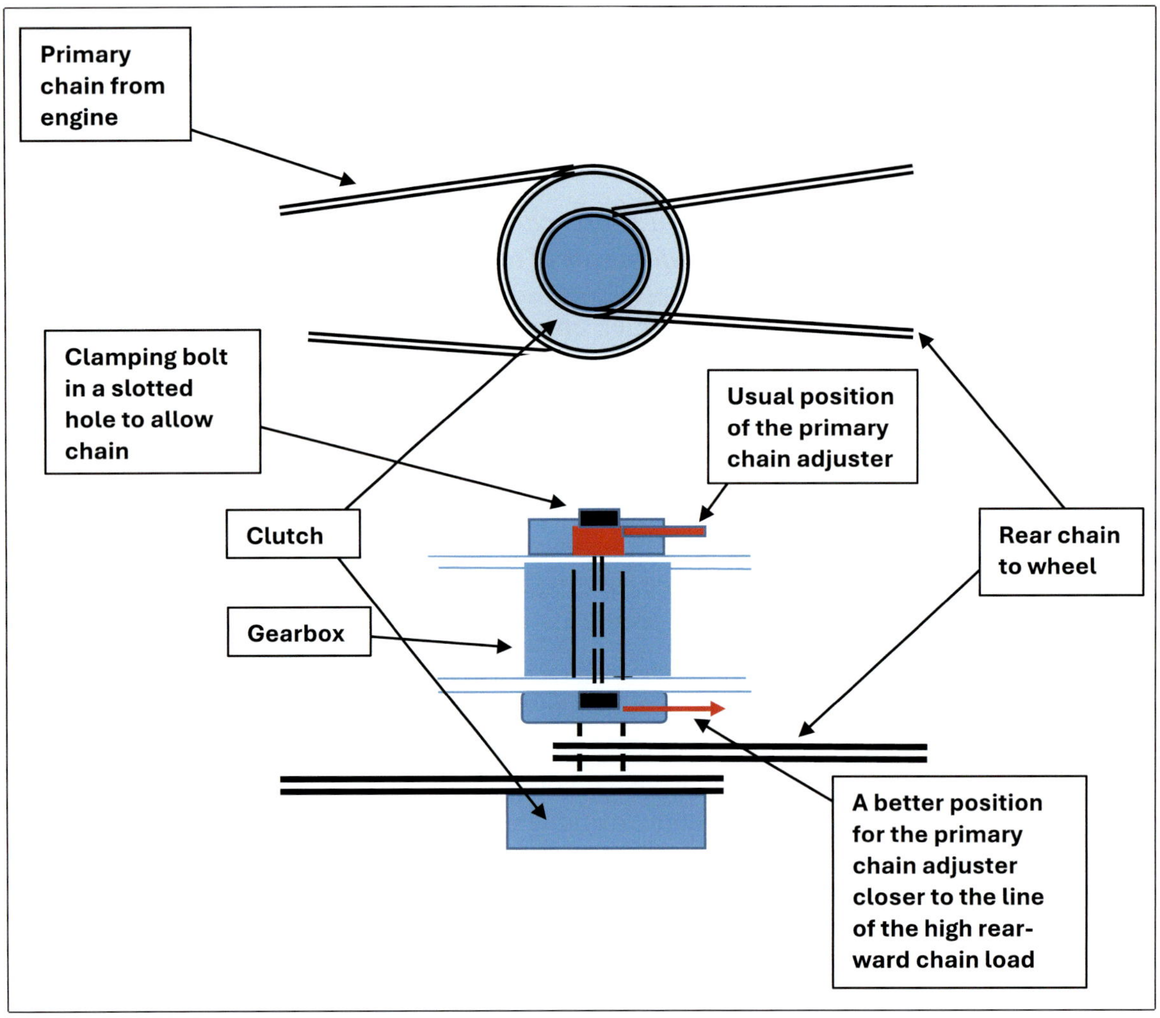

Sketch 21 – Typical classic primary chain adjuster.

A feature of many classic bikes of this type is the positioning of the primary chain adjuster on the opposite side of the bike from the chain, as shown in Sketch 21, presumably to make it more accessible, but this means that it is a long way from the chain line where the chain tension loads are applied. With this arrangement, the clamping of the gearbox in position is the main means of preventing the gearbox being pulled rearwards and any looseness in the clamping bolts could result in overloading of the primary chain which is not likely to be noticed until the chain fails. My answer to this has been, where possible, to reposition the primary chain adjuster as close as possible to the chain line, even if it makes it less accessible. Then the chain adjuster would prevent gearbox movement even if the clamping bolts were loose. (Another example of failure analysis thinking. If a potential problem is identified, no matter how improbable it seems, you should see if anything can be done to avoid the problem before relying on probabilities, in this case, the probability of the clamp bolts becoming loose, to ensure that it does not happen.)

The following check list is included as a guide to the points which require some thought to avoid possible **Hazards** with a chain or belt. Anything which is not directly taken from an equivalent production machine should be reviewed to see if there are any potential **Hazards** associated with it and to justify a low probability of any **Hazardous** failure.

<table>
<tr><th colspan="3">Chain and Belt Drives</th></tr>
<tr><th>Type of failure & possible effects</th><th>Possible causes</th><th>Typical justifications for a satisfactory low probability of failure</th></tr>
<tr><td rowspan="4">Chain or drive belt failure.

Possible rear wheel locking

Possible rider injury

Possible engine/gearbox casing damage causing oil leakage onto rear tyre.

Effects would be Hazardous if either the rear wheel could be locked or if impact with the engine/gearbox casing caused a large oil leak.</td><td>Insufficient clearance between the top run and the chain guard or other static feature</td><td>Is there sufficient clearance for the chain to run off cleanly with no possibility of the wheel being locked by a "balled-up" chain?

A belt drive is unlikely to suffer this effect due to its inherent stiffness preventing "ball-up".</td></tr>
<tr><td>Belt failure</td><td>Is there a top chain guard of sufficient strength to retain the flying end of a broken chain or belt (including an exposed primary chain on a classic bike)?

If not, can one be fitted?</td></tr>
<tr><td>Insufficient clearance between front sprocket and engine/gearbox casings.</td><td>Is there sufficient clearance for the chain to run off cleanly, with a low probability of balling up between the gearbox sprocket and the engine/gearbox casing?

A "ball-up" at this point would stop the engine instantaneously or could break through the engine or gearbox casing causing a large oil leak, which could be Hazardous.

If not, could a smaller gearbox sprocket be fitted with the gearing corrected by a change to the rear wheel sprocket?

Wheel locking from this cause is unlikely as the chain would be free to run off the rear sprocket provided that the bottom run is not enclosed.

A drive belt is unlikely to suffer these effects due to its low weight and inherent stiffness preventing "ball up."</td></tr>
<tr><td>Bottom chain case or similar feature restricting run-off of a broken chain below the rear sprocket</td><td>Is there any possibility of trapping of a broken chain at the rear sprocket?

Any ball-up of the chain between a chain case and sprocket would lock the back wheel.

If there is a possibility of trapping, suggest altering the design to either increased clearance or to ensure that bottom chain case is not strong enough to retain a broken chain.</td></tr>
</table>

8 – Shaft Drive

Shaft drive introduces another gear set into the drive at the rear wheel and the points noted in the gearbox section also apply to the rear wheel bevel box. On most modern systems, the drive shaft is fully enclosed but exposed shafts were often used on older machines. Exposed drive shafts are generally smooth with no projections, but if there is anything rotating which could catch on say loose clothing, luggage straps, etc., for instance a universal joint, it should be guarded to avoid any possibility of something becoming wrapped round the shaft and perhaps pulling rider into the shaft causing injury or loss of control of the bike.

A unique shaft drive system probably falls into the "too difficult" category for any special builder, so a standard production system is probably the only option if you want to incorporate shaft drive. Total failure of a standard production shaft or total failure of the universal joints is very unlikely without any prior signs of deterioration and can probably be discounted as "not reasonably expected to occur" unless there has been a large increase in maximum shaft torque compared with the original production use of the parts.

The following check list is included as a guide to the items which require some thought to avoid possible **Hazards**, although these will vary somewhat dependent on the design.

Shaft Drives		
Type of failure & possible effects	**Possible causes**	**Typical justifications for a satisfactory low probability of failure**
Gear failures in rear bevel box Possible locking of the rear wheel. **Hazardous**	Locking of the final drive gears	Are any of the types of failure listed under gearbox failures applicable to the final drive unit? If so, why is the probability of failure acceptably low?
Contact with the rotating shaft. Possible injury to the rider or loss of control **Possibly Hazardous**	Clothing, luggage straps, etc. caught on projections on rotating shaft and becoming wrapped round the shaft and pulled into it.	Is the shaft fully enclosed? If not, are any projections from the plain shaft, e.g. universal joints, guarded to prevent contact with anything close to them?

9 – Brakes

Legal requirements for brakes always require two independent braking systems on the basis that failure of both at the same time is very unlikely. Unfortunately, on a motorcycle, weight transfer during braking means that the front brake is essential for any emergency stop, so total failure of the front brake at the wrong moment could be **Hazardous**.

The other potential **Hazards** are:

- The front or rear brake locking on and not releasing when the lever or pedal is released.
- Where the brake calliper or drum brake back-plate is prevented from rotation with the wheel by means of an anchor link or similar arrangement, detachment or failure of the link may result in rotation of the calliper or back-plate overloading the brake pipe or cable and, either breaking the pipe or cable leaving zero braking, or by pulling on the cable and locking the wheel.

The best failure analysis justification of a low probability of total loss of braking, available to a special builder, is to ensure that the system used is equivalent to a similar production system which has been shown to be satisfactory by a large number of machines in normal use.

The best justification for preventing the brake calliper or back-plate rotating with the wheel is to provide some sort of secondary stop which would limit rotation of the calliper or brake plate before brake pipes or cables are overloaded.

Most motorcycle braking systems rely on separate front and rear brakes operated by handlebar lever and brake pedal to provide the two independent braking systems, and this arrangement has the advantage that the rider is able to vary the proportion of braking effort between front and rear as required by road conditions and weight transfer effects. Some attempts at coupled brakes operated from the foot pedal have been made from time to time, so this may seem at first sight to be a tempting route for the special builder. Whatever the merits or otherwise of coupled brakes, of which more below, it is always necessary for legal reasons to keep the handlebar lever, arranged so that it can independently operate a brake if the coupled system fails, for instance if all the fluid is lost from the coupled brake's hydraulic brake system. The only practical way to achieve this on a motorcycle is by using two front brake discs and callipers, or two separate calliper piston sets acting on a single disc, with one calliper piston set operated by the handlebar lever and the other by the coupled system from the brake pedal.

Ever since my early experience with the Excelsior, mentioned in Chapter 1, the importance of the integrity of the brake anchor, which prevents the brake plate or calliper rotating with the wheel has been ingrained in my thinking. An example of this is the double-sided drum brake I used on my "Domiracer" replica (Photo 20). The standard brake anchor on the right-hand leg of the Norton forks comprises of a peg on the brake plate which engages in a slot on the right-hand fork slider, a simple effective design with no loose parts which engages automatically when the wheel is fitted (see Sketch 22).

For the double-sided brake, I recreated the peg on both brake plates and used two right-hand fork sliders to provide the slots. As the maximum braking torque is dictated by tyre adhesion rather than the capabilities of the brakes, and was now shared between two brake anchors, all should have been well. Unfortunately, the slots are not symmetrically placed in the fork slider with the normally loaded side of the slot approximately twice the thickness of the unloaded side, see sketch, and when a right-hand slider is used on the left-hand side of the bike, the weaker side of the slot becomes the load bearing side.

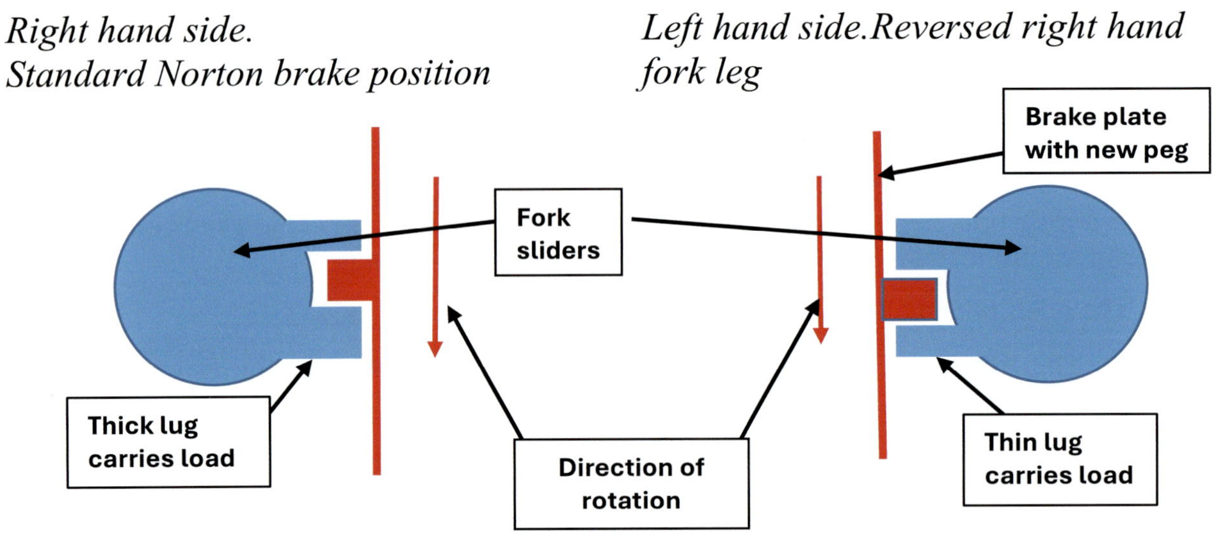

Sketch 22 – The Norton front brake anchor with a double-sided drum brake fitted in place of the standard right-hand side brake.

If you assume that the load on the slot during maximum braking is shared equally between the two sides, the strength of that side of the slot was still sufficient, in theory, but with my normal pessimism when thinking about ***Critical Parts****, I looked at what would happen if the right-handbrake failed to operate for some reason. This showed that the strength of the weaker side of the slot was very marginal if maximum braking was applied with only the left side of the brake working.*

The solution was to provide two secondary brake anchor links attached to two convenient existing stud positions on both the fork sliders as shown in Photo 30. Perhaps overkill, but a confidence booster which weighed very little and guaranteed the integrity of the brake anchor (and did not look out of place).

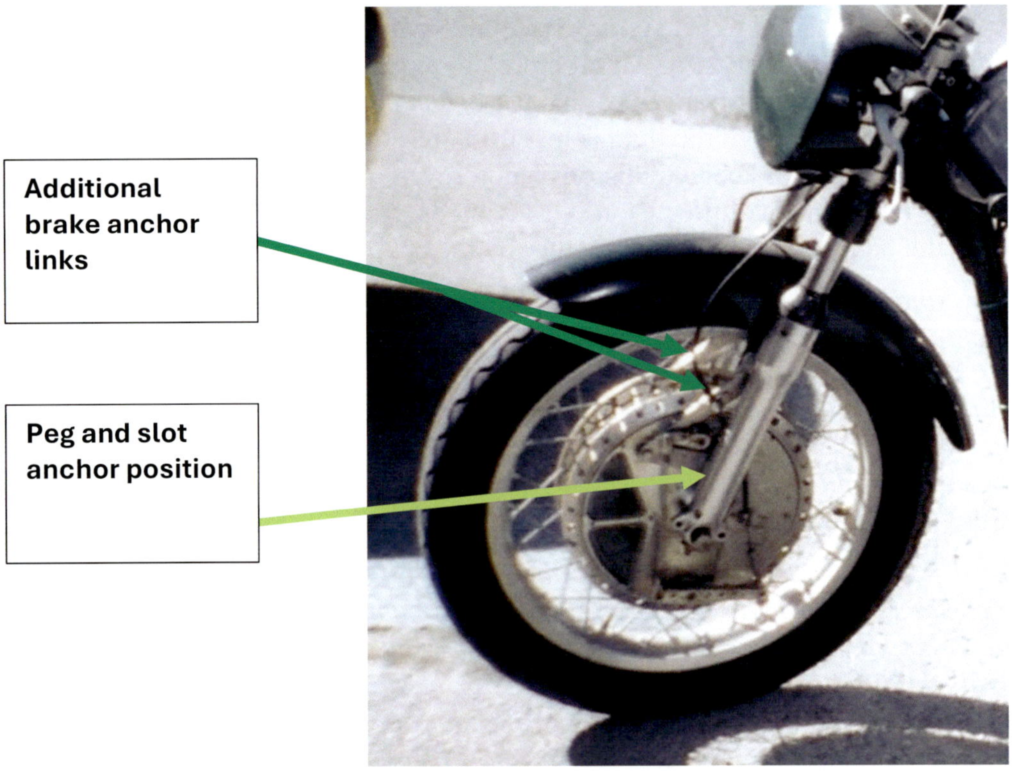

Photo 30 – Double-sided front brake anchors when fitted to Norton forks.

As yet another aside, prompted by the additional brake anchor, it is often near impossible to make two separate attachments, such as the peg and slot and the separate torque link (Photo 30), share the load equally, so that the overall strength may not increase by much. This is because very small dimensional differences mean that one or the other attachment will always become loaded first and take the major proportion of the load. No load can obviously be transferred between two parts which are not touching even if the gap between them is extremely small. However, the duplicated attachments are still safer as, if wear or an overload causes any small deformation of the first support, the dimensional differences will soon be eliminated and the second will take an increasing share of the load.

I am afraid I am wandering on a bit about load sharing but, wandering even further off the subject of brakes — yet another memory from gas turbines has sprung to mind, (nothing to do with motorcycles or brakes, but never mind). I worked on a gas turbine engine for a ship, at one time, which had a power turbine fitted behind the normal turbines which drive the compressors. The power turbine was to be connected by a shaft transmitting many thousands of horsepower to the ship's propulsion system. With no opposing forward load from a compressor to partially balance the rearward gas loads on the power turbine blades and the cooling air pressure on the front face of the turbine disc, this arrangement resulted in a very high rearward load on the power turbine bearing.

This load was too high for a standard ball thrust bearing, so the initial design included two identical bearings fitted one behind the other in the same housing and intended to share the load. However, testing showed that the load share between the two bearings was nowhere near 50/50, in spite of extreme dimensional accuracy, and one or other of the bearings was always lightly loaded and showed skidding marks on the ball tracks. Further testing, using small radioactive pellets inserted into the bearing cages to measure cage speed, showed that the cage speed was much lower than the natural ball rolling speed during acceleration on the lightly loaded bearing, which resulted in the balls skidding.

With this situation, long term durability could not be guaranteed, so the solution was a specially designed single bearing which fitted in the very limited available space. An expensive solution but it worked and there was no more trouble.

The bearings were acting as angular contact ball bearings, which can support both end loads and the radial loads from out of balance and the weight of the rotor, as shown in Sketch 23 below. Loads on the bearing are transmitted along the ball contact line so the angle of this line is critical to load carrying capabilities. Radial clearance and the rolling track profile in the bearing are obviously very important in setting this angle with increased clearance allowing the angle to position itself to align with the direction of the combined radial and axial load, but large clearances also result in deterioration of the other main function of the bearing. That is concentric location of the rotor.

The special bearing referred to above had a non-circular track profile to limit the maximum radial clearance and also give a ball contact angle when loaded axially which provided sufficient axial load capability.

Most motorcycle ball and roller bearings have very little radial clearance as they are intended to carry mostly radial loads, but steering head and wheel bearings also support end loads and it is usually necessary to preload two bearings against one another to provide satisfactory radial location with the necessary radial clearances in the bearings.

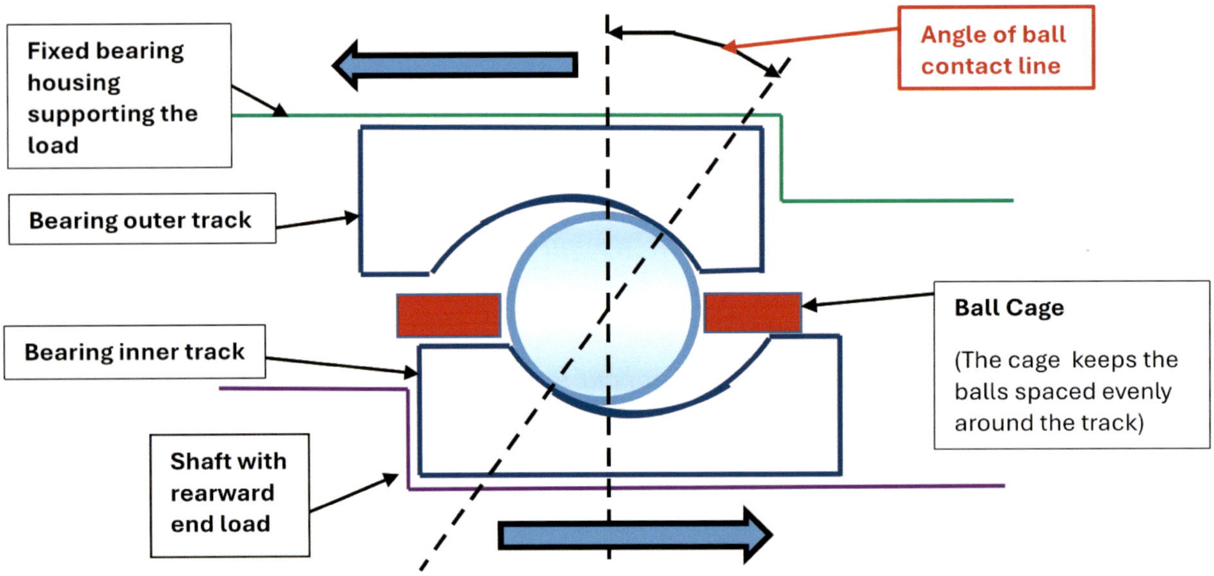

Sketch 23 – Angular contact ball thrust bearing (clearance between balls and tracks exaggerated for clarity).

Time to stop waffling on about load sharing memories and get back to brakes.

Coupled brakes are mentioned above, so perhaps my experience with the system may be of interest. I have never tried to incorporate them in a two-wheeler but have experimented with them on the reverse trike before rejecting the idea. Before you get to the question of braking balance between front and rear, there are two fundamental problems with coupled brakes on a bike.

Firstly, the usual means of operating the brake pedal by pivoting the foot about the footrest on a bike severely limits both the pedal pressure and pedal movement which can be easily applied, compared with a car driving position where the whole foot can move the pedal several inches with pedal pressure reacted against the seat back rest. This is not a problem when the pedal only operates the back brake as a small diameter master cylinder can be used to provide the necessary hydraulic pressure with little pedal load or movement. Operating two or three brakes without increasing pedal movement necessitates master cylinders of at least 2 or 3 times the piston area requiring 2 or 3 times the pedal pressure to give the same hydraulic pressure. The pedal pressure needed for maximum braking on the reverse trike was too high for comfortable operation and the situation would be worse on a two-wheeler as pressure on the brake pedal can only be reacted by rider weight and heavy pedal pressure would tend to unsettle the bike. A brake servo would have solved this problem, but at the expense of added complication and weight and the practical problem of where is there space to fit it?

The second problem is the practical matter of the ability to put a foot down at very low speed or when stopped, which can leave you with no braking from the coupled system. A handlebar lever operated brake is essential to cover this situation, apart from any legal requirements.

After considerable time spent on experimenting with a variety of coupled braking systems for the reverse trike, I have concluded that coupled brakes are not worth including in any special unless they are part of an existing donor bike.

While still on the subject of brakes, the major manufacturers have in general ignored mechanical anti-dive systems for the front brake, preferring to go for higher spring rates to reduce dive and heavier damping of the front suspension, which does not reduce dive but just slows its onset making it feel more manageable. The heavy damping and higher spring rates must also compromise the ideal suspension characteristics for other running unless even more complication with electronically variable damping was included. (You could say that electronic control of damping is a second "sticking plaster solution" to fix a new problem caused by the first "sticking plaster solution", heavier springs and damping, introduced to reduce the effects of a mechanical problem.)

From my experiences with the Velocette and the Horizontal Triumph, there seems to be no significant disadvantages with a mechanical anti-dive system combined with leading link suspension for road use. It is very simple, always works, requires no maintenance, and provides anti-dive effects which are exactly proportional to the rate of deceleration, regardless of rider's weight, brake efficiency or any other variable.

My bikes used a leading link suspension system to achieve the anti-dive effects by angling the brake torque link downwards, with the angle chosen to suit the level of anti-dive required.

It is possible to achieve the same effect with a telescopic fork by pivoting the brake callipers from the front wheel spindle, as is usual on a back brake, and using the system of levers as shown in the sketch to provide the anchor for the calliper. Braking then exerts an upward force on the calliper lever which lifts the steering head through the lever system to counter the downward force resulting from weight transfer. Adjustment of the lever lengths will then allow the level of anti-dive to be set as required. Sketch 24 shows the general idea.

The required lifting force at the steering head to provide 100% anti-dive is about 20% of the braking friction load at the calliper but about 80% anti-dive seems to be the best compromise in my opinion.

The equivalent effect on the same typical bike with a leading link system can be achieved by positioning the calliper at the top of the disc and angling the brake torque link downwards to the fork, an angle of approximately 11° to the horizontal being required for the above example.

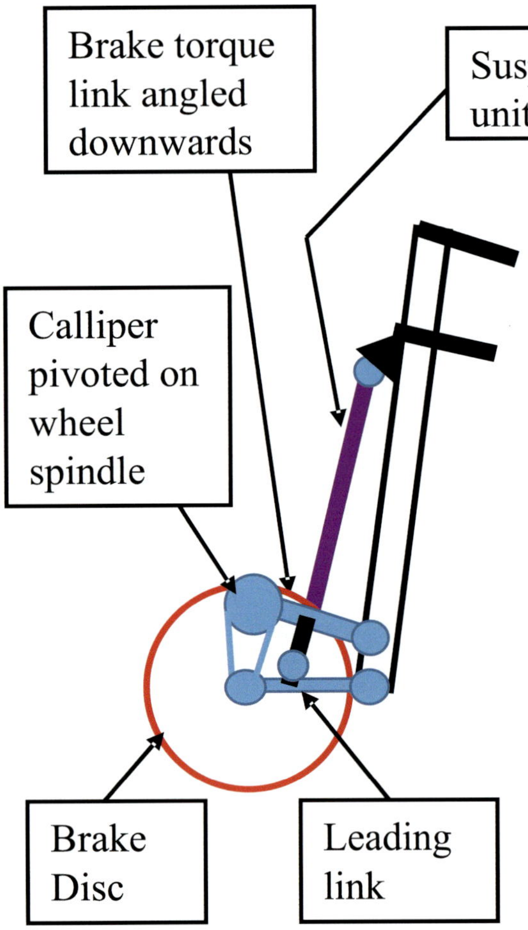

The leading link system

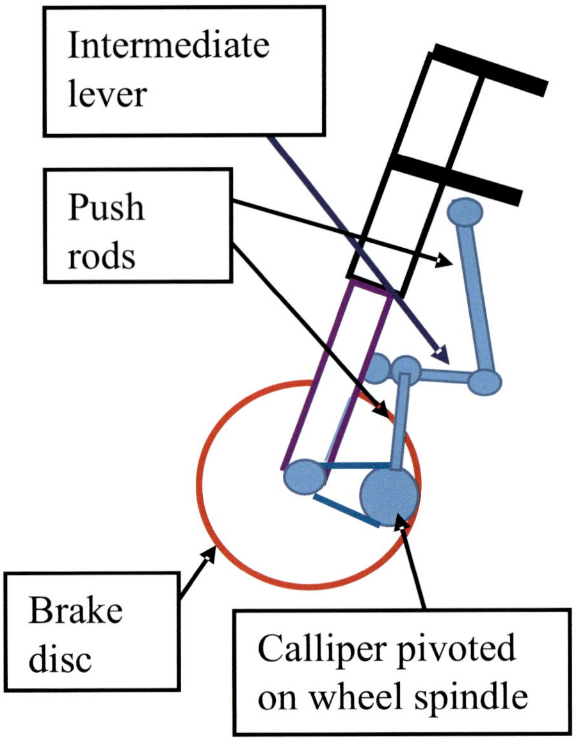

The telescopic system

Sketch 24 – Mechanical anti-dive systems.

I have seen reports that this type of system, when used with telescopic forks on a racing bike, tends to reduce the sensitivity of the suspension movement during heavy braking, perhaps because the calliper moves round the disc during suspension movement, adding an element of friction damping to the suspension. The effect would be to increase compression damping and reduce rebound damping, but I never noticed this effect with the leading link suspension in road use. If this friction damping proved to be a problem, it could perhaps be reduced by adjustment to the hydraulic damping to compensate or by introducing some compliance by using rubber bushes in place of the bearings in the anti-dive linkage.

One bike to use the double lever system with telescopic forks was a variant of the Norton Cosworth Challenge race bike, although they seem to have abandoned the system on later versions. Unfortunately, this bike did not get very far because of lack of finance, although it won some Battle of the Twins races against top class competition.

Before getting back to the main subject, as always, failure analysis would have to be considered with any mechanical anti-dive system. To avoid a range of new ***Critical Parts*** *it would be necessary to provide some sort of secondary stop to prevent the calliper rotating with the disc if the anti-dive linkage became disconnected and to ensure that the push rods in a telescopic system would be retained and would not interfere with the steering if they became disconnected at either end.*

The following table suggests the most likely potential Hazards in a braking system, but all designs are different and the effect of failure of any of the parts should be considered to find all possible Hazardous failures.

Brakes		
Type of failure & possible effects	**Possible causes**	**Typical justifications for a satisfactory low probability of failure**
Loss of Hydraulic fluid The initial effect would be softer feel with longer lever travel followed by zero braking. Given suitable pipes and good quality seals, any seal leakage is most likely to develop progressively and so would probably be noticed before a **Hazardous** situation develops.	Pipe leaks Minor leaks would not be **Hazardous** as they would most likely be detected before total loss of fluid occurred. Sudden total failure of a pipe during heavy braking would be **Hazardous**. Metal pipes in good condition are extremely unlikely to fail but the quality of flexible pipes is less easy to see, so good quality pipes with known history are essential.	Are the pipes standard types in good condition with standard type end fittings? Are they long enough to ensure that they are not stretched or trapped at the extremes of suspension or steering movement (but not too long, which may allow them to become caught on some other feature when the suspension is fully compressed or the steering turned)? If not, suggest changing the pipes to avoid any possible **Hazard**.
	Seal leaks	Are standard cylinders, callipers and seals in good condition used throughout the system, including a totally unblemished surface finish on cylinder bores and pistons at positions where they contact the seals? If not, replace as necessary and check for any leakage.
Drum brake locking on. Inability to release a brake, potentially with a locked wheel would be **Hazardous.**	Excessive lining to drum clearance resulting in operating cams rotating too far. Broken brake shoe return springs particularly on a leading shoe. Any lock up in the external operating linkage resulting in failure to release after maximum braking. A brake flexible cable becoming caught on some other part (e.g. a projecting bolt) during suspension movement, and then pulling the brake on when the suspension returns to the normal position.	Is the cam rotation, from the neutral position to the brake hard on, more than about 30°? If more, reline brakes and preferably machine the assembled linings to give concentricity and minimal clearance to the brake drum. Are the return springs in good condition with no deformation or wear marks at the end locations? If not, replace springs and ensure that there are no sharp edged on the spring location pegs or holes which could wear the springs or result in stress concentrations. Are all lever angles and adjusters positioned so that failure to release a brake after heavy braking cannot occur? If not, revise linkage. Is the brake cable retained in position so that there is no possibility of it becoming caught on anything during suspension movement? If not, secure the cable or preferably remove the projection if possible.

Calliper or drum brake plate anchor (or torque link) failure. **Hazardous** if calliper or back-plate rotation with the wheel could cause wheel locking or brake pipe or cable failure	Lost bolts or anchor location failure.	Could the calliper rotate with the wheel if the anchor became detached? If so, is there a secondary stop which would restrict rotation so that pipes, cables or operating rods are not overloaded?
Failure of disc pad retention system Possible loss of braking. Potentially **Hazardous**	If pad locating parts are not correctly fitted or are in bad condition, pads may work out of position and eventually be lost.	Are pad retention features of a standard production type, in good condition and fitted correctly? If not, repair or replace as necessary. If pad retention is not a standard production arrangement, is it equivalent in integrity? If not, reconsider design
If coupled brakes are used, failure of one wheel brake affecting the other wheel brake. Reduction in braking. Unlikely to be **Hazardous** if a second totally independent brake is available, as is legally required.	If the coupled brakes use a common hydraulic system operated from a common pedal, any failure, such as loss of fluid or linkage failure, may affect both wheels.	Are there separate fluid reservoirs for front and rear brake, and/or an independent additional brake, e.g. a front brake operated by a handlebar lever and connected to a separate calliper? If not, suggest redesign system to include an independent secondary brake. Is the linkage connecting the brake pedal to the master cylinder in good condition with all bolts and pins securely locked? If not, improve as necessary

10 – Cooling System

There are not many **Hazards** associated with cooling systems on a motorcycle, as the worst effect of any cooling failure would be engine overheating which could, at worst, result in engine seizure but a high water temperature warning or loss of power would probably alert the rider before any damage was done. Any modifications which restrict air flow around the radiator, or the cooling fins on an air-cooled engine, should be avoided, remembering that the air needs to flow away after passing through the area being cooled, so a blockage on the way out is as bad as one on the way in.

On a modern machine, the radiator overflow from the pressurised part of the system usually passes into an unpressurised expansion tank which itself has an overflow through a pipe onto the road. Normally nothing will come out of this expansion tank overflow as any water pushed into the tank, due to expansion when hot, will be sucked back into the engine system when the engine cools. Water overflow from the expansion tank is usually due to small cylinder head gasket leaks which allow combustion gasses to enter the cooling system, displacing cooling water. This is not **Hazardous** but could be expensive if allowed to continue.

I did hear of one case where this expansion tank overflow resulted in an accident and was therefore shown to be a potential **Hazard**.

Unusually, the machine had an oil to water oil cooler in the cooling system positioned in the water flow from the radiator back to the engine and the oil cooler section developed an internal leak. The direction of the leak was oil leaking into the water because oil pressure will always be higher than cooling water pressure. The oil would always tend to float to the top in the engine water cooling system and would therefore be the first fluid to be pushed into the expansion tank which eventually overflowed depositing oil on the rear tyre. Not good, but the moral is that all breather and overflow pipes for oil, air or water, should be securely clipped in positions well out of line with the rear tyre and that it is best to investigate any sudden unexpected increase in oil consumption, if you have a water-cooled oil cooler. Don't just top up the oil and carry on.

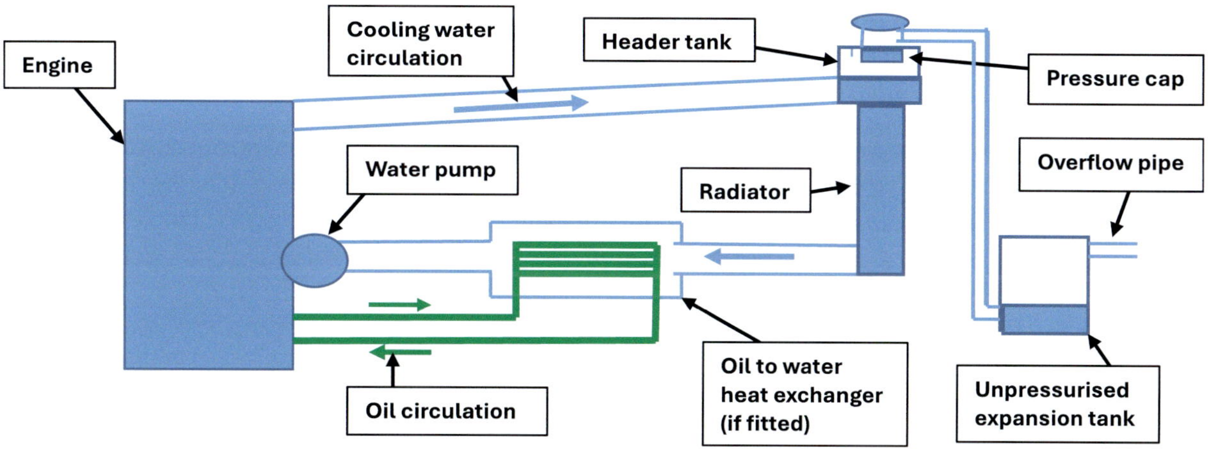

Sketch 25 – Typical water-cooling system with a water-cooled oil cooler.

*Apart from that fairly unlikely **Hazard**, an oil to water oil cooler is a good idea as an alternative to the usual air-cooled oil cooler. It takes less space than an air-cooled oil cooler, can be positioned anywhere, and has the advantage of acting as an oil heater on a cold engine as water temperature will rise more quickly than oil temperature and then it will keep the oil temperature relatively constant at all engine and forward speed conditions. (There are some very compact oil to water heat exchangers used on some cars which may be useful to a special builder.)*

Mention of oil heat exchangers brings to mind a similar problem on aircraft engines.

Aircraft gas turbines often have a very similar system for cooling the engine oil but use fuel as the cooling fluid rather than cooling water. Fuel flows are much higher than in a motorcycle engine, measured in gallons per minute on a large engine rather than per hour, so there is a continuous supply of coolant available to the fuel/oil heat exchanger. There is also a similar failure analysis problem in that oil pressures are always higher than the pressure of the engine inlet fuel, so any internal leak in the cooler would be in the direction of oil loss into the fuel, where it would be burned in the engine unnoticed. The ultimate effect would then be high oil consumption and eventual running out of oil which would necessitate an engine shut down, something to be avoided although not strictly ***Hazardous*** *on a multi-engine aircraft. The strict control of manufacturing quality of the heat exchanger and frequent monitoring of oil consumption contribute to the low probability of total loss of oil in flight.*

The cold fuel does, however, bring with it another interesting problem, related to fuel-cooled oil coolers, in that it is impossible to ensure that there can never be small quantities of water in the very large quantities of fuel used by a large aircraft when refuelling at airports all around the world. This small percentage of water causes no problems in the engine, passing through the combustion section and emerging as very small quantities of steam in the exhaust. The problem arises because, at the end of a long flight at high altitude, the remaining fuel in the tanks will be very cold after several hours in outside air temperatures down to -50°C. It could be as low as -20 to -30°C. The fuel remains liquid at these low temperatures, but any water mixed with the fuel will freeze and will form ice crystals suspended in the fuel. The oil to fuel heat exchanger then becomes important in that it must increase the temperature of the fuel passing through the engine inlet connection sufficiently to ensure that all ice crystals are melted so that they will not build up and block fuel filters or cause any other blockage problem in the fuel system. Much effort is put into ensuring that there is always enough oil heat available to achieve this as it is a problem which would affect all engines equally and could, in the worst case, result in loss of power on all engines simultaneously, which would be ***Hazardous*** *on any aircraft.*

If it is not possible to guarantee enough oil heat in all circumstances, the only alternative is to fit a second heat exchanger using hot air from the engine compressors to heat the fuel. This would add more weight and complication to the engine while, at the same time, reducing engine efficiency slightly by using air which has been compressed but is not then available for combustion and consequently for propulsion.

More memories keep coming back and this one was on a standard mass production car many years ago, related to cooling systems but not in any way ***Hazardous****. There were a number of these cars in the company car fleet, and they had no cooling problems when running but had the unfortunate habit of losing water through the overflow after the engine was stopped immediately after a long fast run. (This was before the days of overflow tanks.)*

The dealers and manufacturer said there was no problem, but it kept happening on more than one car. In the end we solved it ourselves. It was caused by the top water hose which dipped below the top of the engine cooling system before rising to the header tank. This worked well during running as the water pump kept the water circulating but, if the engine was stopped immediately after a hard run with everything hot, enough heat was retained in the heavy cast iron exhaust manifold to boil the coolant close to the manifold and the steam bubble created displaced some liquid water from the top of the cylinder head into the header tank, which overflowed before the steam condensed again. The solution was to add a small vent pipe with an upward slope from the thermostat housing, at the top of the cylinder head to the header tank. There were no further problems. This is another point worth noting if you are altering the cooling system on a special.

This particular problem is perhaps unlikely to occur on a bike because there is generally much less metal in the exhaust manifold to retain sufficient heat, but it is always best to have a continuous upward sloping pipe from the cylinder head to the header tank to avoid any build-up of air or steam in the water passages of the cylinder head. If the radiator is mounted low down with only the header tank and filler cap above the cylinder head, an additional small bore upward sloping pipe should be included.

Sketch 26 below illustrates the problem.

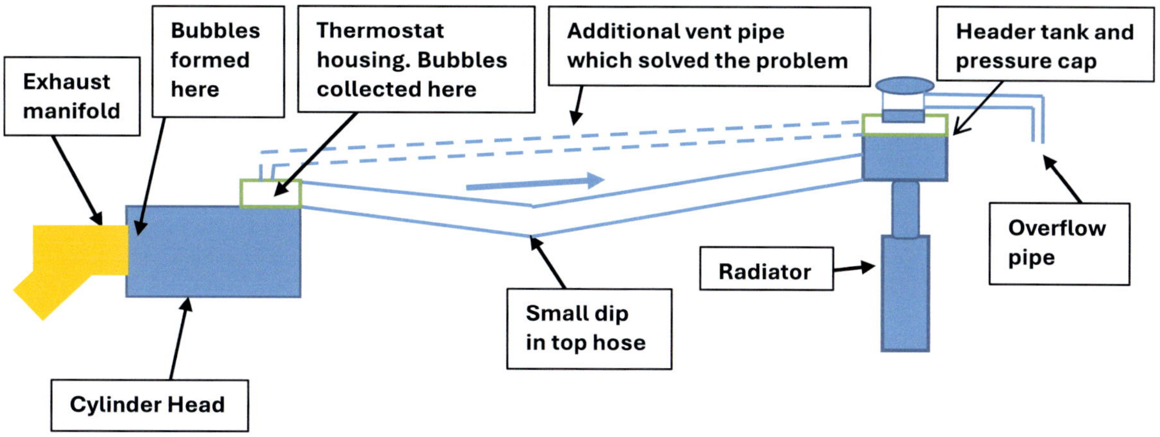

Sketch 26 – Illustrating the reason for water loss following engine stop.

Back to the bikes, another thought on cooling systems is that many current bikes have the pressure cap at the top of the engine pressurised water-cooling system hidden away under the fuel tank where it is impossible to check the water level without removing the tank. This may improve the appearance of the bike but means that the only possible regular check of water contents is at the expansion tank. This check does not really check the water level in the pressurised engine system as any loss through leakage of the pressurised system would not change the level in the overflow tank. On both my reverse trikes, (Photos 24 and 25), I have added small second header tanks. Photo 48 in Part 5 shows the arrangement on the Mk 2.5. These tanks are positioned at the same height as the main header tank and connected to the pressurised system by two small bore pipes, above and below water level. The water level will then be the same in both tanks and can be easily checked in the accessible tank when the engine is cold, and topped up if necessary. It was an easy modification on the trike, with plenty of room to fit the extra tank without spoiling the appearance. It would be more difficult to find a suitable position on a two-wheeler, perhaps behind the headlamp, but I have found it to be a very useful modification well worth considering.

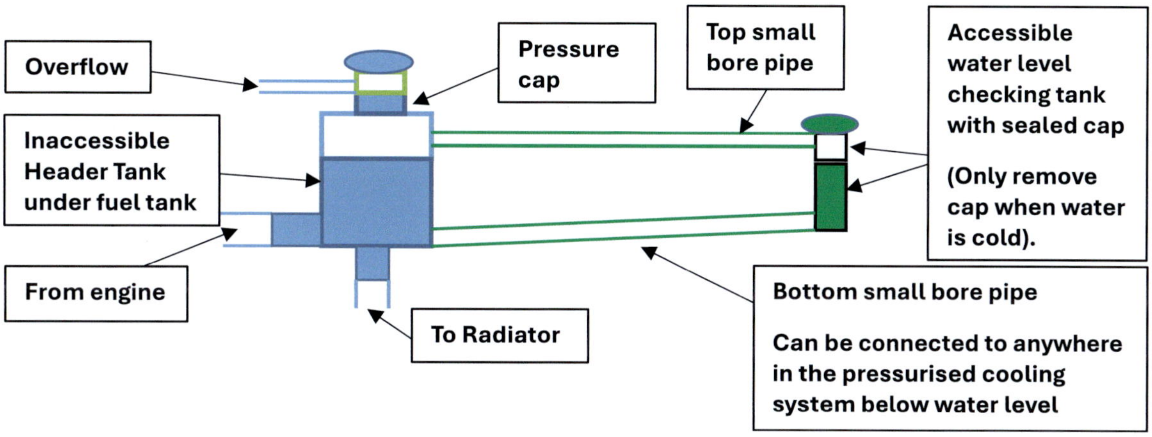

Sketch 27 – Coolant level checking tank (shown green).

The following check list looks at possible failures in the cooling system which could have **Hazardous** consequences.

Cooling System		
Type of failure & possible effects	**Possible causes**	**Typical justifications for a satisfactory low probability of failure**
Water leaks The worst effect of water leakage at any of these points is likely to be engine overheating which is unlikely to be **Hazardous.**	Pipes Radiator core Water pump Cylinder head gasket	Are all parts of good quality and correctly assembled? If not, improve as necessary.
Water loss from overflow. Loss of cooling water on to the road would not be **Hazardous** because the small quantities involved would not significantly affect rear tyre adhesion.	Cylinder head gasket gas leakage may displace water in the system. Water may also leak into the cylinder during cooling with the engine stopped and be lost through the exhaust on start-up. *(A good indication of this is the appearance of the oil which tends to take on a slightly grey, misty colour if it is contaminated with water which is emulsified in the oil during running).*	Are the cylinder head and block faces flat and undamaged with a good quality gasket and correct bolt torques? Leakage may only be detected by water loss with no apparent leaks. *A check on compression pressures when rotated on the starter is a good old-fashioned method of checking sealing of valves and head gasket. (Somewhere around 130 to 150 psi and approximately the same on all cylinders would be good numbers, but if one cylinder is much lower than the others there is probably a problem with that cylinder.)* *(If in doubt, check tightening the cylinder head bolts or restoration of the cylinder head joint faces and a new gasket to prevent future troubles is always a good plan.)*
Internal oil leakage into cooling water at a water-cooled oil cooler, if fitted. Possibly **Hazardous** if oil is lost through expansion tank overflow onto rear tyre.	Leakage at oil/water heat exchanger matrix (The problem with this fairly unlikely failure is that leakage is internal and may not be apparent before it resulted in a potentially **Hazardous** situation)	If a water-cooled oil cooler is fitted, is the water overflow outlet outboard of the rear tyre? This is a failure which would not be **Hazardous** if the cooling water system overflow is routed well away from the rear tyre

11 – External Parts (Fairings, Seats, Mudguards, etc.)

You could say that external parts are all the parts of a motorcycle added to what you may call the basic parts, the frame, engine, transmission, suspension, wheels and controls, to make the complete machine.

In general, external parts would not be expected to fail in a **Hazardous** manner unless they interfered with the safe operation of the basic parts, so the prime necessity is to ensure that all external parts are securely mounted and will not come loose, move, or fall off during running.

Simple precautions will generally be enough to cover these situations such as ensuring that where possible, there is always more than one bolt position attaching a part where movement out of position may affect safety, or if this cannot be arranged, ensuring that the single attachment is strong enough and is securely locked.

Fairings are often the parts with the potential to cause the most problems as they are necessarily of very light construction and could, for instance, move to a position interfering with the handlebars if their front mountings failed or if cracking allowed the fairing to distort due to wind pressure.

The best approach for the special builder is to have a critical look at all external parts and see if any can be moved by hand pressure. If a possibly **Hazardous** point is identified, something can usually be done to improve the design at that position to eliminate the **Hazard.**

There is no real justification for leaving any potential **Hazard** resulting from a likely failure of any external part, and just relying on a judgement of the low probability of it happening, without doing something to improve the situation.

One thing which is always worth doing when making metal, usually aluminium sheet, panels for external parts is to stiffen the edges. This reduces to possibility of cracking in any position with very little weight penalty. The traditional way of stiffening panel edges is to roll the edge to form a circular cross section, sometimes surrounding a stiff wire, but this is difficult to do without special equipment and just folding the edge over by about ¼ inch to give a double thickness or an "L" section is a considerable improvement over a plain cut edge.

The other method is to use rubber or plastic trim strips available in a variety of cross sections from suppliers of material for kit cars, etc. These do not increase stiffness much but can be very effective in damping vibration.

Thin composite fairing panel edges can easily be stiffened by including additional laminates along the edges on the inside of the panel where they will not affect appearance.

Traditional rolled edge (With or without wire) | *Simple folded edge (90 or 180°)* | *Trim strip (Fixed with adhesive)*

Sketch 28 – Stiffening panel edges.

The following table suggests the likely areas where problems may occur, but all designs are different, and everything should be looked at.

External Parts		
Type of failure & possible effects	**Possible causes**	**Typical justifications for a satisfactory low probability of failure**
Loose or lost fasteners. Parts may move out of position affecting safety.	Any failure which makes the attachment of external parts less secure. For example, loss of bolts, rivets or clips, or failure of adhesive bonding, etc.	Is there any position where reliance is placed on a single fastener to prevent potentially **Hazardous** interference with steering or any part of the machine? If so, can additional fasteners be added, or the **Critical** fastener be made totally secure.
Cracking of lightweight items such as fairing panels and brackets. Parts may move out of position, affecting safety.	Fatigue failure due to vibration or aerodynamic flutter effects.	Have you checked with the engine running and when riding to identify any vibrating parts? Is there any stiffening at the edges of thin panels? If not, consider adding at least a rubber or plastic trim strip fixed with adhesive. Vibration which cannot easily be seen can often be detected with the fingertips. If any significant vibration of external parts occurs, can it be reduced by some minor changes to the panel fixing, such as an additional bracket or bolt position?
Potential Hazards?	All machines are different in details such as these, so a check on all external parts will ensure that no unanticipated failure will cause an unfortunate chain of events.	Have you reviewed all external parts to identify any single failure with **Hazardous** potential? If not, check all external parts and modify as necessary to eliminate any **Hazard** found.

12 – Fasteners

The general term "fasteners" covers bolts, nuts, rivets, pins, circlips, and any other type of part which holds things together. Normally, on a motorcycle, there are very few positions where we rely on a single fastener to prevent a **Hazardous** failure but there can be a few so we need to identify them. Perhaps, for example, the rear wheel spindle nut or chain adjuster nuts which if they came loose on some designs could allow the wheel to be pulled out of line, causing the tyre or rim to contact the swinging arm and possibly lock the wheel.

Nuts and bolts, etc. have been around for a long time and their reliability is well established, providing good quality parts of sufficient strength for the job are used. The key to this reliability is satisfactory thread locking and any bolt in a **Critical** position must have a good standard of locking to prevent it coming loose as a result of vibration or alternating loads.

In days gone by, the only means of locking classified as "Class 1 Locking" for aircraft, and thus approved for **Critical** positions, where those which provided a solid metal feature to prevent unscrewing, such as split pins, tab washers, cup washers, and wire locking. These methods are still used today but because they can be a bit inconvenient to use, there is a tendency to rely more on self-locking nuts, spring washers, and thread locking compounds such as Loctite. These are all well proven but are subject to possible unsatisfactory assembly which cannot be seen on the assembled parts. Self–locking nuts and spring washers can lose their locking capabilities if reused too many times and thread locking compounds depend on clean metal threads, with no grease or other contamination, to work properly. Thread locking compounds also have a fairly low maximum temperature capability so they will not be effective on hot parts of the engine or exhaust.

Another diversion, but perhaps of interest, is the potential difficulty of removing bolts locked with thread locking compounds. I would never use them on bolts which did not have good strong "spanning" features such as hexagon or 12-point heads or deep hexagon sockets. Bolts with screwdriver type heads or shallow socket heads can be almost impossible to remove without damage after a few years secured with thread locking compounds. A bit of heat or an impact driver usually does it as a last resort but may be not too good for the paint work. In my opinion, if you are designing a new part which needs threaded holes into a casting or a "made from solid" part, the best solution is to fit threaded studs permanently locked in the holes and then attach the mating parts with self-locking nuts which can be easily removed.

Other points to bear in mind when using bolts in **Critical** locations are:

- In positions where the bolt is loaded in tension, i.e. the loads in use are applied along the length of the bolt, tension put into the bolt due to tightening should exceed the tension load applied during operation. If this is achieved, the bolt will never fail in fatigue as the tensile load in the bolt will never change, only the cause of the tension will alternate between that which results from tightening and that which results from the applied load. A good example of this on a motorcycle is the cylinder head bolts which are loaded in tension by cylinder pressures and, if tightened sufficiently, will maintain sufficient load on the gasket face to ensure sealing even when the maximum cylinder pressure is acting to lift the head.

- In positions where a bolt is loaded in shear, i.e. the forces are pulling sideways on the bolt, the parallel shank part of the bolt which is much stronger than the threaded section, should be long enough to cross the joint plane where the shear load is applied.

- In positions where the shear load is between two surfaces bolted tightly together, most of the shear load will be taken by friction between the surfaces providing that the bolts are tight enough to generate that friction. Again, if surface friction rather than the bolts carries the shear load, the bolts will never fail in fatigue as any remaining alternating loads on them will be low. Wheel spindles are a good example of fasteners loaded in shear at the fork to wheel joint face.

A method I have used for estimating the shear load in a joint which will be entirely absorbed by surface friction, and thus not load the bolt in shear, is to assume a coefficient of friction of 0.1 for dry unlubricated machined metal surfaces, i.e. the surface friction will carry loads up to 10% of the bolt tension due to tightening. Not guaranteed but a good "rule of thumb" method of deciding the size and number of bolts needed for a joint loaded in shear where failure could be ***Hazardous****.*

For the above reasons, correct tightening torque with free running threads, to ensure that the applied bolt tightening torque generates the intended bolt tension and is not absorbed in friction in tight threads, and satisfactory locking, to ensure against loosening due to vibration, are all essential for any bolt used in a **Critical** location.

Bolt length is another point worth considering. As a general rule, bolts used with nuts should protrude at least one thread through the nut when in position. Bolts screwed into threaded holes should engage sufficient depth of thread to ensure that the threads are not overloaded by bolt tightening. A good rule for relatively soft material such as an aluminium casing is a minimum engagement of at least twice the bolt diameter. In stronger materials, a minimum engagement greater than the thickness of the equivalent nut plus 2 threads should be satisfactory although more would be useful if possible, to ensure that if overloaded during tightening, the thread in the casting is protected and only the nut thread or the outer thread on the stud, which can easily be replaced, is damaged.

Other types of fasteners such as rivets, pins and circlips again have been in use for many years and are unlikely to cause any problems providing that standard parts are used, and they are strong enough for the job. Again, the only likely cause of failure is unsatisfactory fitting, such as loose rivets or pins, distorted circlips, or damaged circlip grooves, which may not be easily seen after assembly of the parts. Reuse of locking parts which are bent in position to lock and "un-bent" to release such as split pins or tab washers is a bit suspect and should be avoided if possible, at **Critical** positions.

The other type of fastening which is becoming more common these days, particularly on cars, is adhesive bonding. Again, this method relies entirely on the quality control put into the bonding process and it is often difficult to detect a poor bond until it breaks. For this reason, I suggest that we should avoid bonding as the primary means of holding parts together on our special motorcycle. Bonding is, however, very useful as a secondary method of joining parts which never need to be separated when used in addition to some form of metal fasteners. The bonding material stops any possibility of movement in the joint which could in the long term cause the joint to loosen and excludes water avoiding the possibility of corrosion in the joint.

It is particularly useful on riveted joints and does not need to be a strong bond as the joints' metal fastenings carry the loads.

Even a cheap builder's silicone does a lot to improve the long-term integrity of a riveted joint. I have used this method for many years without any problems and have recently dismantled, with great difficulty, several very old, riveted joints which proved to be as good as new (from parts of the Mk 1 Hardaker car, Photo 25, which were originally assembled in the 1960s, and had been "on the shelf" for many years but eventually came in useful for another job).

One interesting point to be aware of, if using bonding to join thin material, is that the weakest aspect of any bonded joint is peel, where the applied loads are trying to pull the joint apart at one end initially. The greatest load is then applied along a line at the edge of the bond pulling the surfaces apart rather than being evenly distributed over the whole bonded area.

Initial separation of the bonded surfaces at this line then moves the edge of the bond further into the bonded area until the surfaces peel apart completely with much lower total load than the bond would carry if evenly loaded over the full bonded area.

I first came across this in the 1950s on the Canberra aircraft which used a bonding system called Redux bonding, if I remember correctly, to attach some "L" section stiffeners to the inside of aluminium panels. In this case, a single metal rivet at each end of each long stiffener was enough to prevent any failure in peel developing when the panels experienced cyclic loads in flight. Sketch 29 shows the general idea.

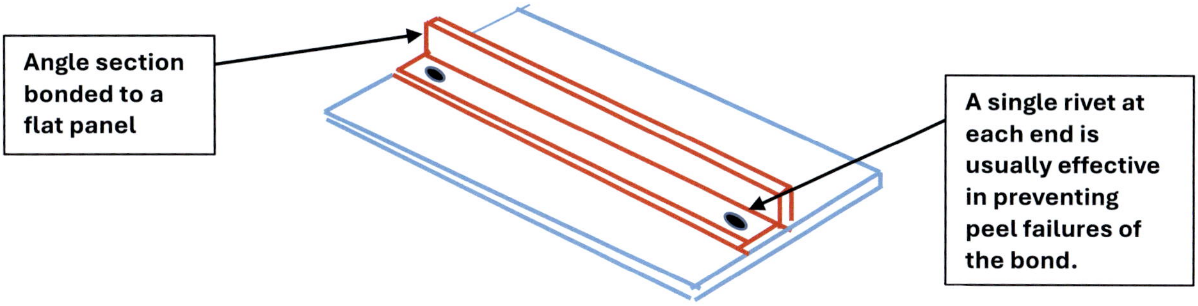

Sketch 29 – Avoiding peel failures of a bonded joint.

The following table suggests identifying all **Critical** fasteners and taking steps to ensure that they will remain secure. All designs are different, and everything should be looked at.

Fasteners		
Type of failure & possible effects	**Possible causes**	**Typical justifications for a satisfactory low probability of failure**
Failure or loss of a **Critical** threaded fastener If the fastener is truly **Critical**, its continued security is essential to avoid a **Hazard**	Lost nuts. Poor thread condition. Incorrect torque on a threaded fastener. Fastener not strong enough for use	Have you identified all single fasteners where loss or failure would be **Hazardous**? If not, check round the machine to identify them (single wheel spindle nuts would be a good start). Are all **Critical** fasteners in good condition and correctly fitted? Is there a satisfactory means of locking on all Critical threaded fasteners? If not, improve or replace as necessary. Where a unique design of fastener not from a similar use on a production machine is used, are you confident of its strength relative to the loads imposed on it? If not, modify as necessary.
Other types of fastener	Unsatisfactory fitting of any other type of metal fastener	Are you confident of the reliability, strength, and correct installation of every **Critical** fastener or **Critical** welded or bonded joint? If not, can the design be improved, or the joint made "non-critical" by providing some alternative load path in case of failure?

Failure of Bonded joints in **Critical** positions Effects of failure depends on position of bond	Insufficient bond strength. Unsatisfactory bonding quality when assembled *(perhaps unsatisfactory surface preparation or incorrect mixing of bonding material or incorrect curing).* Running temperature too high for bonding material *(Very few easily available bonding materials have significant strength at temperatures above 150°C, although some specialist compounds can be effective at higher temperatures.)*	Are there any bonded joints in **Critical** positions? If so, can some additional feature be added to make the joint not **Critical**? How is the quality of the bonding process guaranteed? If not confident of the quality, don't rely on the joint. Is the running temperature of the joint within the capabilities of the bonding material? If not, change the design to either cool the joint or add additional fasteners.

13 – Oil system

The only failures of the engine and gearbox oil systems which are likely to have Hazardous effects are those which will result in significant leakage, which could contaminate the back tyre, reducing adhesion and potentially causing loss of control.

You could argue that total loss of all the gearbox oil could result in a Hazardous lock up of the gearbox and therefore of the rear wheel, but I have never heard of this happening, and I think we can conclude that the probability of it happening is extremely low. In general, the moving parts of a gearbox comprise of gears and ball or roller bearings, both of which will tolerate limited lubrication with no problems other than more rapid wear.

Oil on the tyre is much worse than the other fluids which could leak, water or fuel, as it remains on the tyre and is not rapidly dispersed by further running. Leakage could be from the main engine casings but is much more likely from external pipes, oil coolers or tanks, or oil could be ejected from the engine breather pipe which prevents build-up of pressure in the crank case. Crankshaft and gearbox shaft oil seals could also leak, but quantities of oil lost through these would generally be small. The engine and gearbox oil drain plugs are also possible sources of leakage if not securely locked in position.

It is not possible to conclude that leakage from any one of these positions will never happen, as experience says that it does, so the answer is to make sure that contact with the back tyre is unlikely to result, or where this is not possible, to minimise the possibility of leakage.

Breather and other overboard drain pipes should therefore be securely clipped to exit either behind the rear tyre or as far out from the bike centre line as possible, or led into a catch tank, as required for racing, to retain any oil loss.

If an under-tray is fitted as part of the fairing panels or to provide a guard against impact damage on an off-road bike, it may tend to collect any oil leakage from the engine area. It is therefore a good idea to make the bottom of the under-tray leak proof and shape the rear end of the under-tray so that any oil in the tray will flow to the rear outer corners, rather than leak from a centre line joint, and drip onto the road as far from the centre line as possible, away from the path of the rear tyre.

The following check list looks at possible failures in the oil system which could have Hazardous consequences.

Oil System

Type of failure & possible effects	Possible causes	Typical justifications for a satisfactory low probability of failure
Oil leakage from drains, breathers and overflow pipes.	Over filling could cause excess oil to be ejected from breathers or overflows. Deterioration in piston sealing could increase crank case pressure and eject oil from engine breather. Oil drain down from a separate oil tank could overfill crank case with oil ejected from breather after start-up.	Are all drains, breathers and overflows securely clipped to discharge as far out from the machine centre line as is possible, or behind the rear tyre? If not, consider repositioning the breather and vent outlets or fit a catch tank. *If there is a mudguard or number plate projecting behind the tyre, an outlet can often be neatly concealed behind the bottom of the number plate.*
Oil accumulation in an under-tray. Any build-up of oil could be deposited on the back tyre during a rapid acceleration. **Possibly Hazardous**	Leakage from engine or transmission or associated pipes and coolers	Where would any oil collected in the under-tray drain out onto the road if it overflowed, for instance during acceleration? If in line with rear tyre, suggest altering the under-tray to provide exit paths at the rear outer corners.
Oil leakage from drain plugs	Deterioration, wear, or unsatisfactory tightening of plugs or connections.	Are all parts in good condition and correctly fitted? Are oil cooler pipes sufficiently flexible and suitably retained to avoid vibration? Are they long enough to avoid any possibility of tension between the end fixings? Are all drain plugs to a production standard or, if not, are they securely locked? If not, improve as necessary.
Oil leakage from oil cooler systems		
Oil leakage from engine or gearbox shaft seals		
Other Minor oil leaks Minor leaks are unlikely to be **Hazardous** but could provide a warning before a more serious leak occurs		
Internal leakage in water-cooled oil cooler, if fitted (covered under cooling systems) Could be **Hazardous** if it resulted in oil leakage from the coolant overflow	Possible undetected leakage If undetected, the water cooling system could overflow before the loss of oil became apparent to the rider. Oil, floating on top of the water, would be the first fluid to be deposited on the road and possibly on the rear tyre.	Is the water cooling system overflow pipe outlet positioned as far as practical from the centre line out of line with the rear tyre? If not, reposition outlet.

I have experienced very few major oil leaks on any of my specials, although minor leaks are often difficult to avoid on classic bikes. One common problem on classic bikes, which can result in a significant oil loss, results from oil drain down through the oil feed pump from the separate oil tank to the crank case, with the oil ejected through the breather following start-up, a problem commonly known as "wet sumping." Most modern bikes have the oil contained in a sump at the bottom of the crank case so do not experience this problem.

There are many solutions to this problem, most of which either introduce a manual shut off valve or introduce a valve which is opened by oil pump suction on start-up, in the oil feed line from tank to feed pump. Both have disadvantages.

It is too easy to forget to open a manual valve before start-up, so some form of safety system is necessary, usually an electrical switch linked to the valve which isolates the ignition if the valve is not open. A valve which is automatically opened by oil feed pump suction during start-up, has the disadvantage that the small suction pressures available necessitate very low spring closing pressure making the valve prone to leakage, which defeats the object of the valve.

One alternative which I have fitted on my 1954 Norton Dominator, Photo 16, is to leave the oil feed system unchanged with no attempt to prevent leakage to the crankcase through the feed pump apart from ensuring that the feed pump clearances are not too large. Some oil leakage into the crank case during a lay-up does occur but I have fitted an oil separator in the crankcase breather pipe to remove any oil from the breather airflow and return it to the oil tank.

Then, if the engine is started after a long lay-up, and allowed to run at low speed for a few minutes, any excess oil in the crankcase is returned to the oil tank through both the return pump and the breather pipe with no oil lost out of the breather outlet.

My oil separator is of the type known as a swirl pot, in which the oil/air mixture flows into a circular chamber tangentially at the outer radius and flows around the pot depositing the oil on the outer wall where it runs down the wall and out at the bottom centre position to the oil tank. The air then exits through the top centre position to the overboard breather outlet.

The Dominator used to lose around a quarter of a pint of oil from the breather outlet after say a 3-month lay-up but has never lost a drop in this way over several years since the modification and has suffered no other ill effects. The oil separator, (made from a stainless steel sugar bowl) is fitted out of sight under the seat.

Sketch 30 shows the general idea.

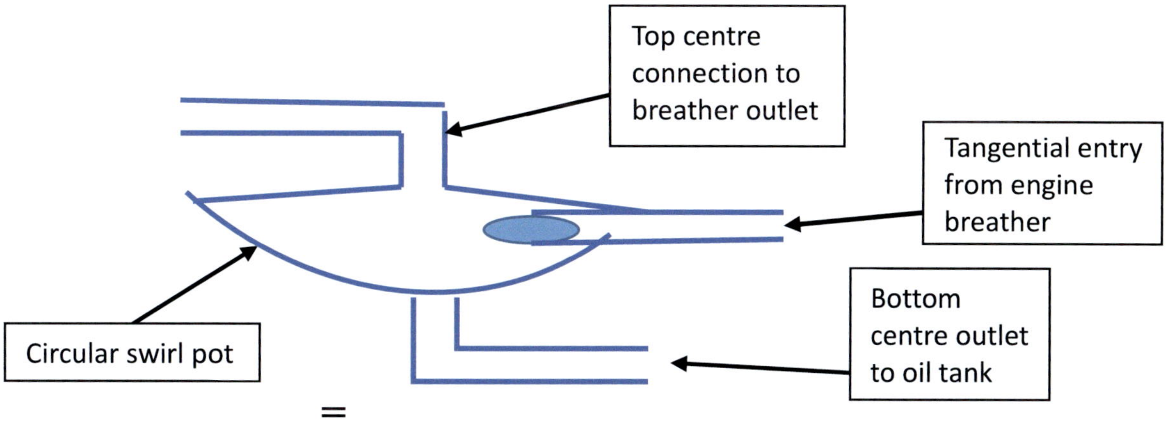

Sketch 30 – Swirl pot de-oiler.
(oil/ air mixture swirls round the outer diameter depositing oil on the walls.
Oil runs to bottom and clean air exits from top centre.)

There are many alternatives to the swirl pot for wet sumping and I have tried some of them.

A gas turbine oil system usually has much in common with a bike with a dry sump oil system in that there is a separate oil tank, and an oil feed pump feeding oil to the bearings and gears. The oil then collects at the bottom of the bearing and gear chambers and is returned to the tank by scavenge oil pumps through an oil cooler and a filter. The tank is often mounted higher than the lowest point in the system so that any oil leaking through the feed pump when the engine is stopped would drain down to the lowest point in the system, usually the bottom of the gearbox.

One method of limiting this drain down on gas turbines, without introducing any restrictive devices in the oil feed, which would be detrimental to the failure analysis, is known as the anti-syphon hole.

(There is also an Airworthiness rule which prohibits any fluid loss onto a runway, so oil leaks when the aircraft is not in flight must be avoided, therefore no system which could allow excess oil to leak overboard could be approved.)

A syphon is a system where gravity flow of liquid from a higher level to a lower level is through a pipe which rises above the higher level before dipping to the lower level. Once started with the pipe full of liquid, this system will flow in exactly the same way as it would if the pipe had a continuous downward slope, but if any air is introduced into the highest point in the pipe, the gravity flow will stop.

In the aero engine system, an upwards "U" bend is included in the oil feed pipe to around the same height as the oil level in the tank. A very small hole in the feed pipe at the top of the "U" bend connects to the oil return pipe from the engine to the tank.

If any drain down occurs when the engine is stopped, the small hole is soon exposed to the air in the empty return pipe and air admitted to the top of the "U" bend which effectively stops any further drain down from the tank at that oil level.

On start-up, the return line is filled initially from return of the small amount of drain down which has occurred and then from recirculating oil. Being submerged in oil again effectively blocks the small interconnecting hole and allows feed pump suction to draw any remaining air bubbles through the system and restart the gravity flow from the tank, mixed with a small amount of return oil drawn in through the interconnecting hole.

Sketch 31 shows the principle.

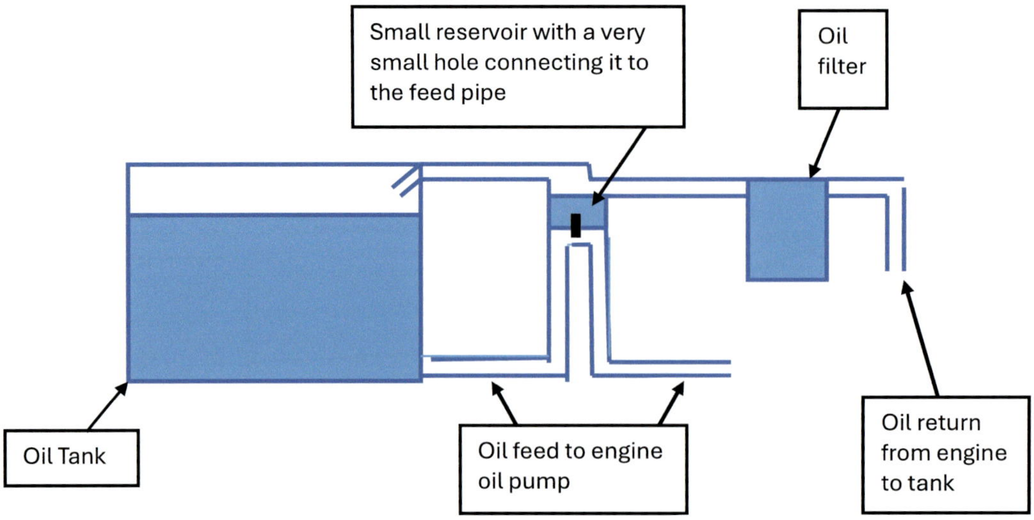

Sketch 31– The anti-syphon hole method of preventing oil drain down to the lowest point in the system when not in use for a long period.

Re-priming would depend on the sealing effectiveness of the feed pump but with only a very small suction pressure, in the order of 1 inch head of oil (1/30 psi or 230 pascals) required, it would be a very inefficient pump if it could not achieve this. An oil pressure gauge or warning light would be a sensible addition for peace of mind although unlikely to be necessary.

Blockage of the interconnecting hole is unlikely if it is positioned downstream of an oil filter, in the oil return line and therefore only exposed to filtered oil. The only effect of blockage would be increased oil drain down when the engine was stopped but, on a bike system, it would be a good precaution to provide an access plug to allow the hole to be easily cleared.

I have tested the idea on the Modernised Dominator using a "U" bend arrangement produced by means of drilled holes in a single aluminium block to be positioned inside the oil tank, (see Photo 31.1 below). It seems to work well, on the basis of limited testing, but would be difficult to accommodate on a standard bike. Photo 31.2 shows the arrangement made from drillings in a solid block of aluminium.

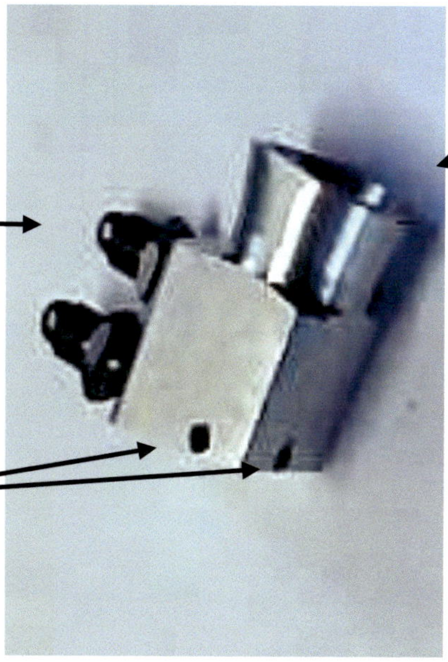

Photo 31.1 – Anti-syphon hole system to be tested on the modernised Dominator

Since then, an even better idea has been suggested for single cylinder engines and 360° twins, which would be much easier to fit to a standard bike with minimum visible change.

If the slow drain down of oil into the crankcase is allowed to occur and the oil is quickly returned to the tank on start-up with no overboard loss through the crankcase breather, the problem would be resolved.

This can be achieved by moving the crankcase breather connection to the bottom of the chamber just above the return pump inlet and blanking off the original breather connection. This bottom inlet is then connected to the top of the oil tank through a light one-way valve with a separate oil tank vent, (a reed type valve would be best as rapid cycling with low opening pressure is needed).

The system uses the crankcase pressure resulting from the descending pistons during starting to drive excess oil back to the tank through the one-way valve which prevents the oil being drawn back as the pistons rise. Once the engine is running, the average crankcase pressure will tend to stabilise at something below atmospheric but there will still be a net outflow through the breather as a result of gas leakage past the piston rings, which will carry any remaining oil back to the tank. The breather and the return pump operation will then return to normal. Sketch 32 shows the general idea. I have made the parts and initial tests of the system on the Modernised Dominator are promising.

Sketch 32 and Photo 31.2 show the general idea and the bottom breather entry in the crank case base plug.

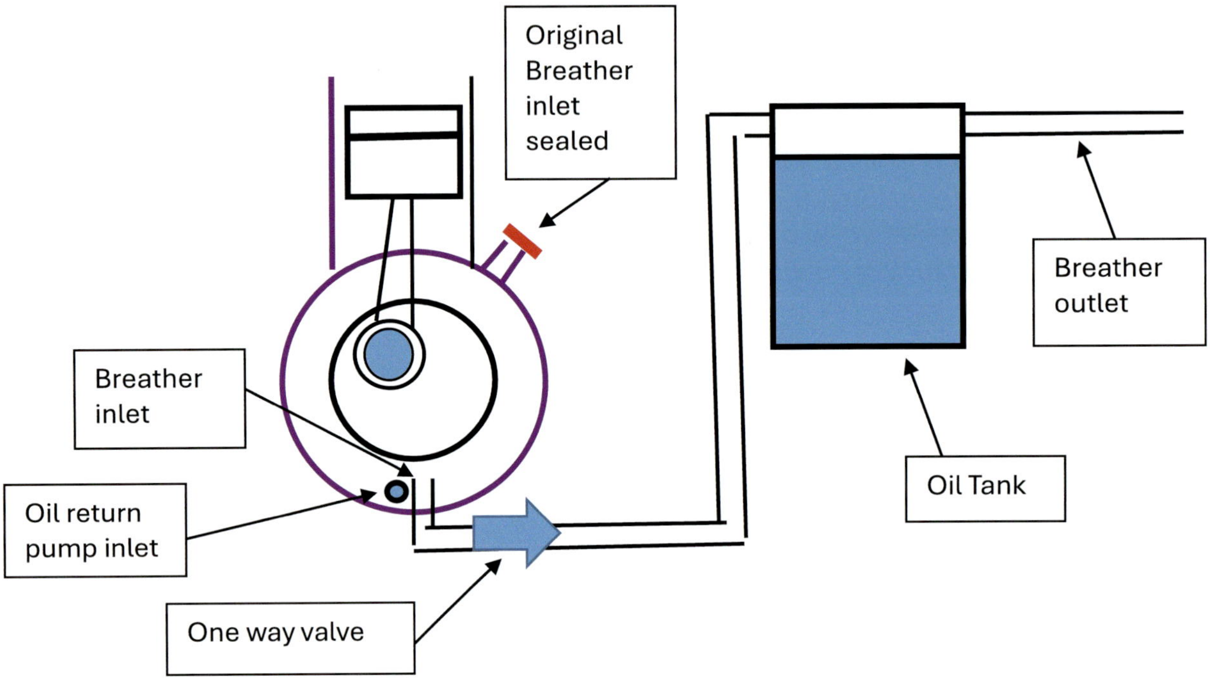

Sketch 32 – Bottom breather arrangement.

Photo 31.2 – Bottom breather entry made for a Norton Dominator.

The breather entry is through the pipe which is above the return pump suction pipe entry, through a gauze strainer in the cylindrical section, but has good clearance from the central flywheel.

The above thoughts have not much to do with potential **Hazards** but may be of interest to classic bike owners and prompt someone to try one of the remedies suggested or perhaps think of something better.

14 – Fuel System, Including the Possibility of Fire

There are not many potential **Hazards** associated with a motorcycle fuel system, but the most obvious is the possible consequences of a fuel leak from the tank, pipes or other fuel system components. Leaking fuel in itself is not likely to be **Hazardous** unless it is ignited by some means and a sustained fire results. Failures of the fuel control system controlling the fuel/air mixture supplied to the engine, whether mechanical or electronic, are covered separately in sections 24 and 26.

Failures resulting in fires which could be **Hazardous** on aircraft are usually associated with leaking flammable fluids, fuel, engine oil, hydraulic oil, etc. or with electrical systems overheating for some reason. There has also been at least one case of lithium-ion battery over-heating and causing a serious fire, fortunately on the ground, which prompted a change in the airworthiness requirements, treating battery compartments as fire zones as described below, to reduce the potential for **Hazards** resulting from this cause.

Apart from a battery fire which is unlikely on a motorcycle while in use, a sustained fire can only occur if there is flammable material, a source of ignition and sufficient air supply for combustion. For the fire to stabilise, the local air speed must also be fairly low as a high air speed, faster than the flame propagation speed, will blow the flame downstream away from the source of combustion.

The general principle of protection against fire in aircraft is to isolate the areas, where the above conditions could occur in the event of a failure, from the rest of the aircraft by means of fireproof barriers (fire walls), and then to provide systems and procedures for controlling any possible fire within these fire zones. These systems include fire detection and extinguisher systems, and some independent means of cutting off the supply of flammable fluids, principally fuel, to the area within the fire zone. A lot of effort is put into identifying possible causes of fire and designing to minimise the probability and potential effects of any fire situation.

For example, an aircraft engine has fuel and oil systems which could leak, hot metal surfaces and electrical systems which could provide sources for ignition and there is usually a supply of air to sustain combustion. Each engine is required to be contained within its own fire zone and provided with detection and extinguishing equipment and an independent fuel shut off valve. The engine operating instructions would then include procedures to be followed by the pilot if a fire is detected which, when deployed, are intended to extinguish the fire before the fire wall is breached.

Not much of this aircraft method is directly applicable to motorcycles but the general principles apply. That is, flammable material, an air supply and a source of ignition are necessary for a fire to occur. The only likely flammable material is leaking fuel, as the engine oil will not be hot enough to ignite easily. The possible sources of ignition are hot exhaust or perhaps sparking or overheating in the electrical system.

On a naked machine, the airflow when moving at reasonable speed is likely to be too fast for a sustained fire without continuous re-ignition, except perhaps in small, sheltered pockets, but a fire under a fairing would seem to be possible, particularly if there is some fibrous sound damping material on the inside of the fairing which has become soaked in oil or fuel.

It would therefore be a good plan to keep all sound damping well away from the exhaust or other sources of ignition. A sustained fire under the fairing burning the rider's leg would be somewhat **Hazardous** at high speed in the middle lane of the motorway.

The most likely time for a fire is when stopped or after a crash when leaking fuel, hot exhaust and an unrestricted air supply may come together. Although expensive, such a fire is unlikely to be **Hazardous**, providing that the rider can get away from the bike. It is, however, always a good plan to keep all fuel system parts within the frame with no parts projecting to positions where they could be impacted in an accident. Another small point is that, if a fuel tank breather pipe is fitted, it should run from the top of the tank to an outlet on the centre line and below the bottom of the tank. This will minimise the possibility of fuel loss through the pipe with the bike on its side or upside down. (I have seen a plastic breather pipe run down the centre of a hollow steering stem, which seemed like a good idea, providing that the outlet is low enough to be below the bottom of the tank.) Careful positioning of the breather outlet is, however, a waste of time if the tank cap leaks.

Another aspect of fire protection is a build-up of flammable gas in an enclosed space which could create an explosive mixture needing only a spark to cause ignition. On an aircraft there is a possibility of a very small fuel leak onto a hot surface in the area between the engine and the surrounding cowling causing a build-up of combustible vapour from evaporation of the leaked fuel and, for this reason, the certification rules specify that this space must be continuously ventilated by means of a controlled small airflow providing something like 2 or 3 air changes per minute in the whole volume of the cowling airspace to prevent any build-up of an explosive mixture.

There is no real equivalent to this type of enclosure where an explosive mixture could accumulate on a motorcycle, so there is no danger of this type of explosion. Fairings, for example, will always have sufficient airflow through them to carry away any fuel vapour.

*The only case of fuel vapour explosion I ever heard of on a motorcycle was on a classic bike with an enclosed primary chain case which also enclosed the alternator. The seals were blown out of the chain case by explosive pressure on more than one occasion and it was eventually concluded that a sufficient quantity of petrol vapour, passing the pistons into the crank case, had leaked into the chain case to form an explosive mixture and ignition had been provided by an electrical fault in the alternator connections. A very unusual problem, but not likely to be **Hazardous.***

Another aircraft problem is the avoidance of build-up of static electricity in any metal part which is not earthed to the main structure of the aircraft, such as a pipe which is both mounted on rubber and connected into the system by rubber connections. If enough static charge could be built up, it could cause a spark discharge which would be a potential source of ignition. The certification rules require that any such part should be earthed to the main engine structure, which is itself earthed to the airframe, by means of additional bonding leads if necessary. This bonding is also essential in the event of a lightning strike, which happens quite often in flight. If all conducting parts are electrically connected to each other, the possibility of sparking between parts or disturbance of electrical and electronic systems is much reduced.

There is very little chance of either build-up of static or a lightning strike on a motorcycle, but a totally isolated metal pipe could become charged to battery voltage, without blowing a fuse or showing up in any other way, if it was in shorting contact with an electrical wire, just waiting to become a source of ignition at the wrong moment. The answer is to ensure that all metal pipes in the fuel system, or anywhere else, are electrically connected to the frame and that fuel pipes and electrical wiring are not in contact anywhere on the machine.

The following check list looks at possible failures in the fuel system which could have **Hazardous** consequences.

Fuel System, Including the Possibility of Fire		
Type of failure & possible effects	**Possible causes**	**Typical justifications for a satisfactory low probability of failure**
Fuel leakage from pipes, connections, and fuel system units. Possible fire if leaking fuel can come into contact with a source of ignition. Could be **Hazardous** in some circumstances.	Deterioration of non-metallic material (rubber or plastic pipes, connectors, seals, or other fuel containing parts). Unsatisfactory connections Vibration of unsupported metal pipes resulting in metal fatigue.	Are all pipes and connections in good condition and not deteriorated due to ageing? If not, repair or replace as necessary. Are all pipe connections of a type in normal use and correctly assembled? If not, repair or replace as necessary. Are all pipes suitably supported such that they are flexibly mounted and under no inbuilt strain? Are they sufficiently restrained to prevent excessive movement in vibration? If not, modify as necessary.
Electrical sparking between metal pipes and items earthed to the frame. Possible fire if leaking fuel can come into contact with a source of ignition. Could be **Hazardous** in some circumstances.	Contact between metal pipes and electrical wiring with the possibility of insulation damage.	Are any electrical wires in contact with metal fuel pipes? If so, re-route pipes or wiring. Are all metal parts of the fuel system electrically connected to the frame or other battery earth? If not, add some sort of earth connection.
Electrical fuel pump. Continued operation after a downstream leak. Large amounts of fuel would spray from the leak, increasing the risk of fire. Could be **Hazardous** in some circumstances.	With the machine upright, the leak may not stop until the ignition is switched off. The rider would be in control at this time and would be able to switch off. Following an accident, the rider would not be in control and the leak could continue, most likely resulting in fire.	If the machine uses an electrical fuel pump, is there an automatic electrical shut off which prevents fuel flow when the engine is stopped, even with the ignition still on? If not, consider adding a suitable safety system. Does the engine emergency stop switch also switch off the fuel pump, either directly or as a result of cutting the ignition? If not, could this be arranged? (Some machines use a special relay in the power supply to the pump which is activated by the ignition system when the engine is turning, but cuts off when rotation stops.) *(The snag with this system, as I have found on several carburetted machines, is that if the bike is left standing for a period or runs out of fuel, the carburettor float bowls will not refill without "endless" cranking with the starter.* *On two of my bikes, I have fitted priming switches, which are spring loaded off, and bypass the cut off relay, when held in the on position. This allows the pump to be operated with the engine stopped but does not interfere with the safety system at any other time.)*

NEW FUEL TYPES

While on the subject of fuels, the moves towards low carbon emissions are resulting in fuels containing increasing quantities of ethanol which is a plant based fuel and therefore counts as renewable. Currently E5 and E10 fuels include 5 and 10% ethanol respectively.

How this may develop in the future is suggested in Part 4.but needs to be a consideration in starting any new project.

*I cannot see any likely additional safety **Hazards** with E5 or E10 fuel but there are some reliability issues when considering classics or anything more than a few years old.*

The effects on engine running are not likely to be significant for road use but, as ethanol includes some oxygen in its chemical makeup, it needs proportionally less oxygen from the air for complete combustion. That is, it needs a marginally richer mixture, about 2% richer for E10, and is likely to use 2% more fuel by volume for the same running. A more modern engine, with electronic fuel control including an exhaust gas sensor, should automatically make the correction, and the difference is sufficiently small to have little effect in normal road use on an older machine, but it is something to watch for the future.

It is worth remembering that there is nothing new in the use of alcohols in fuel. Methanol, (another alcohol with a larger percentage of oxygen in its chemical composition), was used in racing fuels in the 1930s as it allowed higher compression ratios than the available petrol but was prohibited in most forms of racing post war, although it continued to be used in speedway.

The main effects of ethanol on older machines occur over a longer period, particularly if they are not used regularly, and tends to sit with the same fuel in the system. Ethanol has two basic problems in these circumstances,

1) *It absorbs water from the atmosphere and will mix with any proportion of water, unlike fossil fuels where any water content tends to collect separately at the bottom of the tank.*
2) *It attacks many plastic and rubber compounds causing swelling, disintegration, and separation of bonded joints.*

A combination of these two unfortunate effects can cause rusting inside fuel tanks, leakage from fuel, pipes and seals or blockage of carburettor jets and passage-ways, or other similar problems.

The only good answer is to replace everything suspected of lack of ethanol resistance with equivalents in modern materials and coating the inside of fuel tanks with some ethanol resistant coating, but these things are not always possible on older machines.

In that case, the best alternative is to try not to leave the machine for long periods with fuel in the tank and system, particularly in damp conditions. There are fuel additives available, which are advertised as providing protection, and are worth a try. Millers Vespa Power Plus, Frost Ethmoid and Flexolight Ethanolmate are approved by the Federation of British Historic Vehicle Clubs, but there are many others.

*I am trying Lucas Safe Guard, which seems to recommended, and will see how it goes with the confidence that the worst that could happen is a failure to start, which may be inconvenient but is unlikely to be **Hazardous**.*

15 – Mechanical and Hydraulic Controls

Motorcycle controls comprising of throttle, brakes, clutch, and gear change, twist grip, levers, cables, hydraulics, etc, all using conventional mechanical systems, are well developed and unlikely to cause any **Hazard** apart from the possibilities of front brake cable or front brake hydraulic failure or a stuck open throttle at the wrong moment. An acceptably low probability of these **Hazards** can be ensured by sensible design using standard parts, or equivalent one-off parts of equal strength, and ensuring satisfactory manufacture based on the long experience on production machines.

Most failures of hydraulically operated brakes result from gradual deterioration, minor fluid leakage needing frequent topping up, or impending calliper seizure causing the brake to drag or be slow to release after braking.

Internal leakage of the master cylinder seals or air bubbles in the fluid may not cause fluid loss but would result in a soft feel to the lever and so be obvious to the rider before any total brake failure could occur. The exception would be sudden loss of fluid from the front brake, most likely from failure of the pipe.

Standard metal or flexible pipes rarely fail unless damaged, so protection of the pipe from rubbing on other parts, vibrating, or being bent or stretched during running are the best ways to avoid a pipe failure.

This is particularly important where the front or rear brake pipe crosses from the fixed part of the forks or frame to the moving part of the suspension system where the full range of movement should be checked to ensure that the pipe does not become stretched or kinked or rub against some other part during the movement.

If a cable operated front brake is used, there are several possible **Hazards** to consider apart from cable failure. The brake plate end of the cable needs to be securely attached to the lever and the lever to the brake cam shaft. The cable outer must also be securely fixed in a suitable strong anchor point. All these attachments should be locked to prevent loss of nuts, pins, etc., due to vibration.

The routing of all cables and brake pipes should also be checked to ensure that they cannot be trapped or stretched or caught on other projecting parts by the extremes of steering and suspension movement and that cables and pipes are not in direct contact with electrical wiring. The area around the steering head is a place where cable trapping is likely to occur and should be checked with all cables, electrical wiring, and pipes in position, as it is often a very congested area where trapping can occur.

I have seen one case where the bundle of electrical wires from the handlebars became trapped between the steering lock stops, at the bottom fork yoke position on the steering head, severely restricting movement of the steering. Fortunately, it only resulted in the bike on its side at low speed but it was a somewhat expensive way to discover the necessity of ensuring that all cables and wires passing round the steering head are secured in a safe position.

Straying into electrical systems covered in the next section, an engine stop switch, or use of the clutch, are the last resort answers to a mechanical problem causing a stuck open throttle, but a stop switch is useless in an emergency if it cannot be operated easily with both hands on the handlebar grips. The stop switch allows continued engine braking which would not be available if the clutch was used to cut power to the rear wheel and also prevents the engine from running up to high speed. The stop switch should, preferably, stay off when operated, not return to the on position when released, as many do on older bikes.

With the throttle stuck open approaching a slow corner, you are likely to be rather busy. Trying to brake, lining up for the corner and checking for other traffic while holding the stop switch down to prevent the still rotating engine from re-starting could be somewhat challenging.

The usual arrangement of a stop switch is for it to either cut power to the ignition system, or to earth the ignition coil so that the current through it is constant and no sparks will be generated.

*Off the subject again, but on a similar subject, in spite of endless regulation and type approval legislation, many new production cars still have the horn button in the steering wheel boss where it would be impossible to reach while steering to avoid an unexpected **Hazard** (provided that the driver actually sees the approaching **Hazard** while playing with all the built-in distractions in a modern car).*

Cable length often needs to be altered to suit unique machines with the position of the end fittings, usually the cylindrical nipple end at the handlebar, altered and the nipple re-attached to a new longer or a shorter cable by solder or brazing. Care is therefore essential to achieve a satisfactory joint particularly on a front brake or throttle cable.

Failure of a front brake cable which has been altered in length, by the cable pulling out of the re-attached end fitting or breaking close to the reattached end, is probably the most likely **Critical** part failure which could occur on a bike with cable operated drum brakes. It is a good example of the reason for careful control of all manufacturing processes on a **Critical** part.

Many modern cables have cast on end fittings which will need to be cut off to shorten the cable and reattach new end fittings by more conventional means. The easy method of confirming that the joint is sound is by an overload test, for example by pulling the brake lever with a lot more force than would ever be used for braking and checking that there has been no movement in the joint or damage in any other part of the system.

Brazing cable end fittings onto the cable using low temperature braze (silver solder) will give a stronger joint than soft soldering and is essential if the inner cable is of stainless steel, but with a conventional (old-fashioned) inner cable I prefer soft solder as it has been proved over many years to be strong enough. It is too easy to overheat the cable close to the end fitting when brazing which could reduce cable strength and make the overall result weaker.

The traditional method is to separate individual strands at the ends of a cable, thoroughly cleaned and then splayed out after passing through the nipple before soldering. This will ensure good solder contact with all surfaces. Any protrusions can then be trimmed to the nipple profile after soldering. The expanded end produced will resist cable pull through even if solder strength or its adhesion to the nipple is not ideal.

Sketch 33 overleaf shows the general idea.

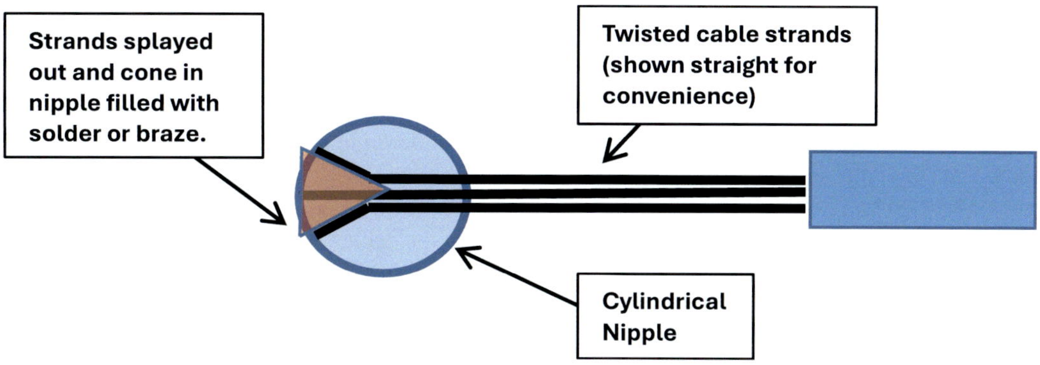

Sketch 33 – Cable end fitting in a cylindrical nipple. The individual cable strands are spread into a cone within the conical hole in the nipple and filled with solder or braze to form a solid cone which cannot pull through the nipple even if the solder adhesion is not the best.

The following table shows the types of failure of mechanical and hydraulic controls which could be **Hazardous**. All parts should be reviewed as all designs are different and it is often the small details which influence the outcome of any failure of this sort of system.

Mechanical and Hydraulic Controls		
Type of failure & possible effects	**Possible causes**	**Typical justifications for a satisfactory low probability of failure**
Hydraulic fluid leakage. Gradual leakage is unlikely to be **Hazardous** as it would be detected by wetness on the brake parts or a soft lever or pedal before total brake failure. Sudden loss of fluid from the back brake is also unlikely to be **Hazardous**. Sudden loss of fluid from the front brake is potentially **Hazardous** as it is most likely to occur during heavy braking when you really need the brakes.	Poor quality flexible or metal pipes and connections. Seal failure in master cylinder or calliper. Damage to pipes either due to unsatisfactory clipping or road debris.	Are the master cylinder, pipes, connections and calipers all standard production types in good condition? If not, are you confident of their integrity? Are the pipes suitably held in position so that they are not stretched and do not foul on anything during full suspension and steering movement? If not, improve the design.
Metal pipes in contact with electrical cables which could cause damage in case of a short circuit. An unlikely failure but it would not be impossible for a short circuit to burn through a pipe, which could be **Hazardous**.	Rubbing together as a result of vibration could wear insulation. In most circuits this would have no worse effect than blowing a fuse, but in a high current circuit a substantial short could occur.	Are any hydraulic pipes in contact with electrical cables? If so, suggest move them and secure pipes and cables separately.

Control cable failure. Failure of the front brake cable could be **Hazardous** as it is likely to fail during heavy braking when you really need the brake. Failure of the throttle cable is unlikely to be **Hazardous** if the throttle closes automatically but if the failure holds the throttle open, it could be **Hazardous** in some circumstances.	Poor quality cables. Sharp bends cause most strand breakages and are most likely to occur close to the nipple ends. Lack of lubrication making the cable stiff to operate and possibly reluctant to return to the initial position. Cables caught on other projections stretched during steering or suspension movement.	Are the cables of good quality with cable ends secure? (Modern production cables often have permanently cast on nipples which rarely cause any problems, but earlier or modified cables will have nipples brazed or soldered in position.) If not, improve as necessary. Are the cables smooth to operate and arranged to return rapidly to their initial position when released? If not, replace cables. Have you checked that the cable length and positioning is satisfactory at all positions of the steering and suspension? If not, check that the cables are not pulled tight, or severely kinked and cannot get caught on any projection, even if deflected by wind pressure or for any other reason.
Throttle stuck open. Likely to be **Hazardous** if the engine power cannot be controlled separately by provision of a stop switch	Stuck throttle cable. Broken return springs.	Is there an engine stop switch in a position where it can be operated without taking hands off the handlebars and which does not return to on position when released? If not, suggest providing one. *Systems with separate opening and closing cables are inherently safer than those with spring closure only, because of lower cable loads in normal operation and the ability to apply a high closing load through the closing cable in an emergency. (The closing cable is usually very lightly loaded, due to spring assistance, and is unlikely to fail.)*
Cable operated brakes. Detached cable to lever connections. Lever to brake camshaft connection. Detached brake cable outer stop. Any of the above could be **Hazardous** if operation of the front brake cable is affected	Inadequate locking of nuts or pins resulting in loosening and loss in vibration conditions.	Are all fasteners suitably locked to prevent loss? If not, improve integrity and/or locking. An occasional overload test when stationary by means of a good pull on the lever is not a bad idea.

16 – Electrical System and Basic Electronics

In general, there are very few potential **Hazards** associated with what you may think of as the basic electrical system on a motorcycle, that is, the generator, battery, lights, ignition, horn, etc. The only obvious ones are as a source of ignition for some other failure, e.g. a fuel leak, or sudden loss of headlights at the wrong moment.

The possibility of providing a source of ignition for a fire can be much reduced by separation of electrical wiring from any component carrying flammable fluid, i.e. fuel or oil, and by positioning all wiring and electrical components within the frame so that they are unlikely to be damaged in an accident, e.g. by contact with the road.

The possibility of headlight failure is usually covered by the use of separate electrical feeds through different fuses to main and dipped headlamp beams so that some light can be restored with the dip switch in case of a failure at the wrong moment in dark conditions. Failure of the headlight earth connection could also be **Hazardous** if it occurred in dark conditions as both main and dip beams would then be cut off.

Failure of the earth connection to the light unit is very unlikely if the light unit is mounted directly onto the metal forks, but many modern machines have the light unit mounted on non-conducting composite fairing parts and rely on wiring for the earth connection from the lights. This arrangement is much more prone to damage or failure than a direct connection through the forks and frame, and failure of a single wire could leave you in sudden total darkness when approaching a corner on a dark night. In any new or modified wiring system, duplication of this negative connection to the frame to two separate points on the light unit and on the frame would be a very good idea.

Failure of the taillight is unlikely to be **Hazardous** as reflectors and glow from the headlight would usually show the machine's position to other road users, but the failure may not be noticed immediately so efforts should be made to minimise the possibility of failure. The use of a light unit with more than one bulb in both rear and stop light positions is a good move unless maintaining original appearance on a classic is important.

Control of the throttle in some cases is entirely electrical these days, with no mechanical cable connection, so the possibility of failure to close would need to be covered. As with a cable operated throttle, as mentioned in the previous section, an independent stop switch cutting the ignition would seem to be essential as it is hard to see how any electrical or electronic system can be more reliable than a well-maintained cable.

Double insulation of electrical wiring using insulating sleeves or conduits of insulating material at points where rubbing and contact could occur is a good plan, particularly around the steering head where the wires are frequently flexed by moving the steering. It is a good idea, if possible, to run pipes and cables on one side of the steering head and wires on the other. The separation of electrical wiring from brake pipes or fuel pipes at other positions, particularly metal pipes, is also a good plan to guard against loss of electrical insulation and a short circuit through the pipe possibly causing a brake fluid or fuel leak and a possible fire.

If designing a unique electrical system, the fuse arrangement should be based on good failure analysis principles such that no single fuse failure could result in a dangerous situation.

Similarly, there are few potential **Hazards** associated with what you may call the older generation of motorcycle electronics, that is, control of ignition, fuelling, and handlebar instruments (clocks) and warnings. Most potential failures will only result in, at worst, loss of, or no, engine power or a flat battery. Inconvenient and perhaps costly but not **Hazardous**.

You would hope that any loss of electrical power to an electronic fuel control unit or between the unit and the fuel injectors would shut down the engine but, for this, reliance must be placed on the system manufacturers. A good reason for not interfering with this type of system or the power supplies to it.

The following table indicates the sort of things to look for in reviewing the electrical system but, as always, all parts should be checked to ensure than nothing has been overlooked. The possibility of an electrical short circuit providing a source of ignition for a fire is covered under fuel systems, but is obviously something to be aware of in the failure analysis of any electrical system.

Electrical System and Basic Electronics		
Type of failure & possible effects	**Possible causes**	**Typical justifications for a satisfactory low probability of failure**
Sources of ignition Not normally **Hazardous** unless a fuel leak occurs simultaneously.	Worn insulation allowing short circuits to the frame or other earthed components. Bad connections or breaks in wiring causing sparking.	Is all wiring aligned into suitable bundles and double insulated where in contact with earthed components? If not, improve lay out as necessary. Is all wiring supported by clipping or cable ties to fixed parts such that it can- not vibrate or move into contact with hot or rotating parts? If not, improve layout as necessary. Are all connections secure and with good electrical contact which is not lost if the wire is pulled or flexed? If not, improve connections.
Stop switch failure. The stop switch is a safety system and failure could be **Hazardous** if throttle control of the engine failed.	Bad connections or switch failure.	Do you use the stop switch for stopping the engine after normal running? If not, suggest testing the switch operation occasionally to ensure that it has not become disconnected. Is the stop switch easy to operate in an emergency while holding both handlebar grips to maintain control of the bike? If not, move it to a better position.
Stop switch reverts to the on position when released. This could be a dormant **Hazard** as holding the switch in the off position to prevent the engine restarting may be difficult in an emergency.	Type of switch fitted.	Does the stop switch stay in the off position when released? If not, suggest changing switch for one which stays off.

Headlight failure Potentially **Hazardous** if it failed at speed in dark conditions.	Fuse, bulb, switch or wiring failure. Headlamp unit earth connection failure.	Is the dipped beam wired separately from the main beam through a different fuse. If not, suggest change system to allow light to be restored using the dip switch in case of main beam failure. Is the earth connection of the headlight unit, usually through the headlight mounting bolts, secure? If the light unit is mounted in a non-conducting composite or plastic fairing, is there a secure separate earth lead to the frame? If not, improve earth connection and preferably add a second independent connection.
Taillight failure. This would not normally be **Hazardous** as reflectors, etc. would still show the position of the machine to following traffic.	Fuse, bulb, switch wiring or earth connection failure.	Is all wiring securely clipped in position with secure connections? If not, improve as necessary. Suggest checking the route of the earth connection, which is often through numerous parts and joints, to ensure continuity in conditions of vibration, etc. *(There is usually no other way of checking that the rear light is working, other than looking behind, so a failure during a run may not be noticed immediately. All possible efforts should therefore be made to ensure that the system is secure and, if possible, a light unit with more than one bulb should be fitted.*

Just another thought on the design of unique electrical wiring systems for a special when no ready-made production parts will do the job.

All basic motorcycle electrical systems are fundamentally the same, comprising of a generator feeding a battery which supplies numerous electrical units, such as lights and horn, and the various electrical subsystems, such as the ignition, indicators, and electronic units, with fuses and switches where necessary in each circuit. Each unit then requires an earth return connection to the battery. Most modern bikes use a negative earth return but the older classic bikes often use a positive earth which can make the fitting of modern units a bit difficult.

Sketch 34 (next page) shows this basic layout in simplified form.

The circuit diagrams shown in overhaul manuals and similar publications are often a bit confusing as they tend to combine the basic circuit with the layout of wiring on the bike and use a large number of symbols which mean that you have to keep going to a list on another page to see what the symbols mean, then you lose your position on the cluttered diagram and have to start again.

If you are designing your own system, I find that the best way is to start with something like Sketch 34 to collect together all the things which need to be connected in an easy to follow format with the top horizontal line representing battery positive and the bottom line negative.

Then apply a bit of Failure Analysis type thinking to decide where a single wire and a single fuse could supply more than one piece of electrical equipment and where an independent supply through a separate fuse would be safer.

There are not likely to be many potential ***Hazards*** *with the electrical system, other than those mentioned in this section, but it is easy to list the things you would least like to happen,* *such as sudden failure of the headlights when approaching a fast corner at speed on a dark country road. This sort of equipment needs to have a secure independent electrical supply, and preferably a plan to deal with a failure thought out in advance.*

The electrical supply to other items, where failure would not have significant safety effects, can then be combined to operate from the same supply wiring and fuses.

The next step is to decide which equipment will only be connected if the ignition switch is on and which needs to remain connected with ignition off. For instance, a two-wheeled bike can have almost everything turned off by the ignition switch, although the radiator fan is often left on so that it can continue to cool a stationary engine until switched off by a thermal switch in the cooling system. On a trike, it is legally required in the UK to have sidelights and a hazard flasher system which can operate with ignition off.

The third step is to arrange the routing of the system on the bike so that all the wires going in the same direction can be bound together in bundles or sleeving to allow secure clipping of the bundles and ensure that they cannot interfere with steering, etc. The routing of a unique system is best achieved by actual assembly in position on the bike as wire lengths, etc. are very difficult to specify in advance.

I have often used the 7-core wire sold for caravan and trailer connections as a neat way to provide several wires running in the same direction. Two 7-cores will usually provide most or all of the wires needed for connections passing round the steering head and will provide a neat and flexible system.

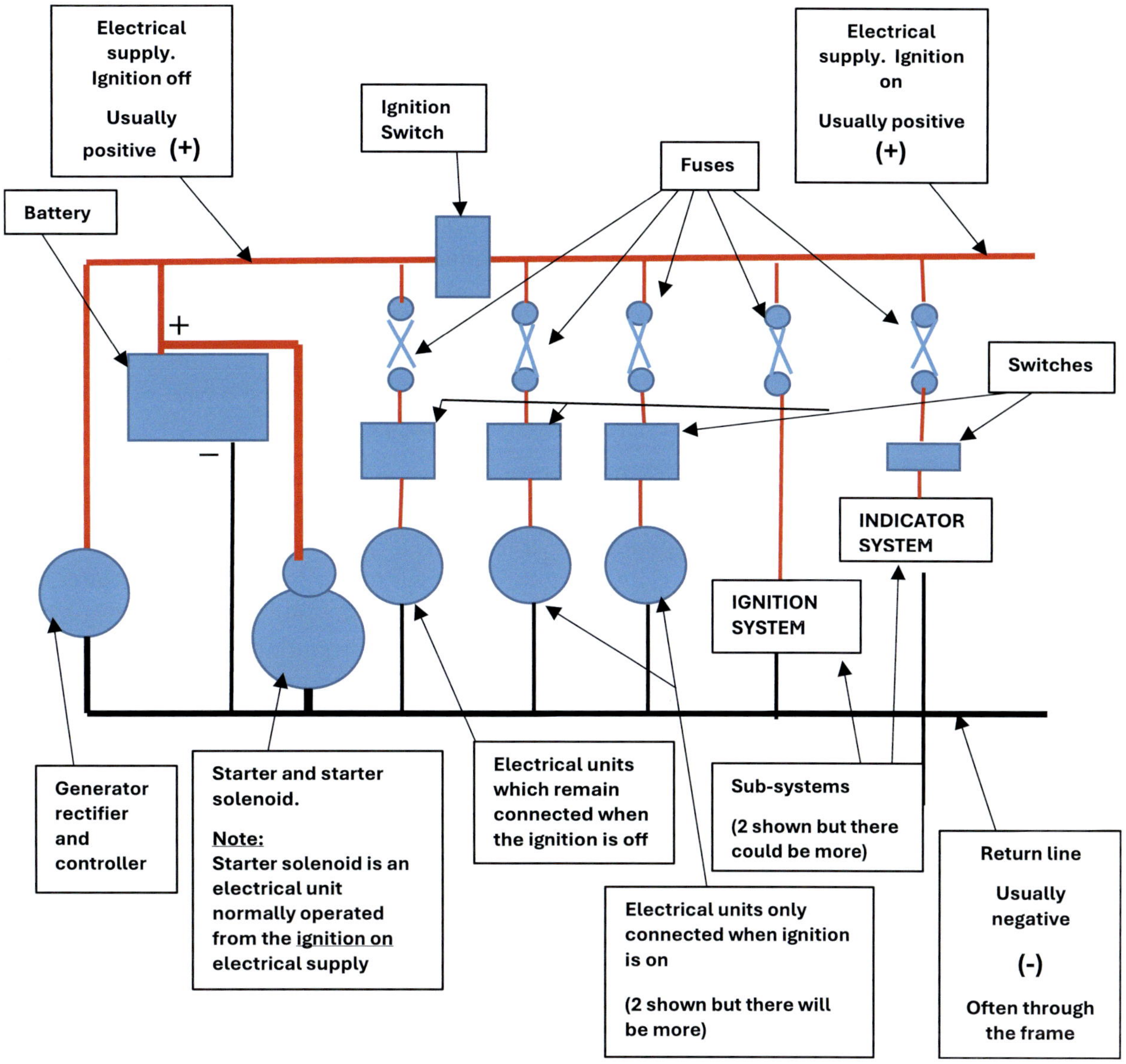

Sketch 34 – Simplified basic electrical system.

A similar simplified circuit diagram can then be drawn for each of the subsystems, such as the ignition system and the turn indicator system.

17 – Electronic Rider Aids

The new generation of electronics, e.g. ABS, traction control and electronic control of engine power in response to a combination of twist grip position, traction control and engine management computer programming, introduce a new range of potential **Hazards** which would be difficult, if not impossible, for the special builder to deal with unless he uses a complete unaltered system from a major manufacturer. The manufacturers will have spent a large number of man hours covering all possible eventualities and should be able to quote a satisfactory record of operation of the system, both in their own testing and in general use by the public. Some adjustments to system settings are possible, for instance minor changes to fuel scheduling in response to dynamometer testing, but this sort of change does not alter the basic safety logic of the system.

The basic reasoning behind the adoption of these systems is that both engine power and braking power of modern large capacity bikes has increased much faster than tyre adhesion has improved with improved rubber compounds, etc. It has therefore become much easier to exceed the limits of tyre adhesion with small changes in throttle opening or brake lever pressure. This naturally results in the rider being more cautious in approaching the limits. The electronic rider aids effectively limit the power of both engine and brakes to match available tyre adhesion more closely, thus giving the rider more confidence that he is not going to suddenly spin up or lock the wheels with a small twitch of the throttle or brake lever.

To illustrate this effect, a simple calculation for an average bike based on the likely maximum weight on the back wheel during acceleration and the maximum likely coefficient of friction between a road tyre and a normal dry road surface shows that it is physically impossible to transmit more than about 450 to 550 lb ft (610 to 746 newton metres) of torque at the rear wheel during acceleration, depending on the tyre rolling diameter, (and often a lot less on a slippery surface), before the limit of traction is reached and the wheel spins up. This is equivalent to an engine torque of about 45 to 55 lb ft, with a 10 to 1 overall gear ratio, or 70 to 80 horsepower (bhp) (52 to 60 kW) at the wheel at 60 mph.

The rear wheel torque differences due to different rolling diameters do not affect the horsepower figure as a smaller wheel rotates faster at the same road speed. and so will transmit the same amount of power as a larger wheel.

The other limitation to the maximum torque which can be transmitted is the front wheel lifting, as a result of acceleration, which may occur before the wheel spins up depending on the road surface conditions and the weight distribution of the bike. This would reduce the maximum transmittable torque as power would have to be reduced, either with the throttle or by absorbing some power with the back brake, to prevent the whole bike rotating backwards around the wheel.

The effect of weight transfer during acceleration is easily shown by the fact that, if a high powered stationary bike is held on the front brake to prevent acceleration, a burst of power to the rear wheel will inevitably spin the wheel up, but if the same power is applied with the bike free to accelerate there will be much less wheel spin and more tendency to lift the front.

The maximum power which can be transmitted in normal dry road conditions, all other things being equal, will increase in direct proportion to increasing road speed, to something like 130 horsepower (97 kW) at 100 mph, but if we are talking about road bikes, even higher speeds are not very relevant.

For comparison, the power necessary to maintain a steady speed of 100 mph is usually somewhere between 30 and 50 bhp (22 to 37 kW) depending mainly on the aerodynamic drag of the bike. (A "travelling barn door" will require a lot more power to push it along than one with a small frontal area and an aerodynamic fairing.) Any available excess rear wheel torque, over that required to maintain the constant speed, is then available for acceleration.

It follows that, in normal road-going conditions on a very high-powered bike, it is never possible to use the engine's maximum power at full throttle and high engine revs in a low gear for acceleration at normal road-going speeds. Some form of power restriction, whether by delicate manual throttle control or by electronic means, is therefore essential.

Where the very high-powered bike gains over an 80 bhp bike in road-going conditions is in the availability of the maximum transmittable torque over a wide range of engine speeds, that is mid-range torque. A smaller engine would only develop its maximum torque over a small range of engine speeds and would need more frequent gear changing to keep the engine in the speed range for maximum torque during acceleration, to achieve its maximum performance.

Racing slick tyres with the best rubber compounds on good track surfaces have more adhesion than road tyres as illustrated by the high lean angles mentioned in section 3. These lean angles indicate about a 40% increase in adhesion compared with the maximum I have assumed (perhaps optimistically) for road tyres. If this increase also applies to traction, it would mean that the transmittable torque at the rear wheel and equivalent horsepower would also increase by 40%. At road-going speeds, this would still be lower than the maximum engine power, approaching 200 bhp, which is claimed for some superbikes.

The best practical demonstration of acceleration in the road-going speed range is provided by the times achieved in ¼ miles sprints from a standing start. Classic sprint machines on modern tyres, with maximum power much less than 100 bhp, can achieve times, with a skilful rider, in the same region as those achieved by modern production sports bikes with over 150 bhp and all the modern electronic aids, showing that tyre grip and keeping the front wheel down are the limiting factors, not maximum engine power.

(My Vincent once recorded a 13.2 second standing ¼ mile, a respectable time for a modern bike, with not much more than 50 bhp but with good torque over a large speed range and the Norton clutch modification which allowed much better control of rear wheel torque during the start than the standard Vincent clutch. The stopwatch timing used in those days would not have been as accurate as modern timing systems but wouldn't have been that far out.)

The above is a long diversion from the subject of electronic controls, perhaps stating the obvious again, but in order to list possible **Hazards** with a system, it is always first necessary to be clear as to what the system is intended to do.

As I see it, with admittedly little practical experience of these systems, *(too old and cautious to intentionally go anywhere near the limits of adhesion these days),* they are undoubtedly a safety aid for road use, if kept in reserve for emergencies, where they should prevent the rider making matters worse by inexpert use of throttle or brakes. This is demonstrated by the long-term use of ABS systems on road cars and increasingly bikes, where the number of times the system is called into operation is limited to the occasional emergency.

The new **Hazards** come when the use of, and reliance on, the systems become routine, as it could easily be, particularly on high-powered sports machines, for example lack of caution using throttle or brakes on slippery surfaces and relying on the system to sort it out. The **Hazards** may then come to light on the one occasion, perhaps in the whole life of the bike, when the system fails to operate as intended. This could be due to a minor wiring or sensor fault or some such problem which is not detected while still upright and in control, or even just a small delay between the sensor input and the system response.

Total reliance on these systems is generally agreed to be acceptable in racing, as they allow the rider to concentrate on the most important objective, getting to the finish first, and the acceptable level of risk is perhaps somewhat higher when racing. Also, race bikes generally spend more time being maintained, tested and perfected than they do racing, so deterioration as a result of high mileage, corrosion, exposure to water and road salt, wear and tear, etc. is unlikely.

The potential **Hazardous** situations which may result from failures in these new generation electronic control systems are easy to see and comprise mainly of:

- Failure of the electrical power supply to the system.
- Failures which result in loss of control of the throttle, i.e. more power than is demanded by the twist grip angle or by correct traction control operation.
- Inability to close the throttle or shut the engine down.
- Too much or too little braking relative to lever pressure or as required for correct ABS operation.

A further point which may be overlooked is the potential effects of overheating or exposure to fire on electronic circuits and control units. In general, electronic components have a low tolerance to overheating and can fail in unpredictable ways if overheated, sometimes at temperatures as low as 100°C. To avoid any possible **Hazard** from this cause, all systems must be fitted in positions where there is sufficient cooling and no significant chance of being exposed to high temperatures.

These points are covered in the table below but the difficulty, from a failure analysis point of view, with any electronic system is that there are always many ways in which it could go wrong and often only one way in which it will operate as intended. Investigating all the possible variations of faults on a system, which is **Critical** to rider safety, is definitely not a job for amateurs and cannot be easy for major manufacturers.

Perhaps I am old-fashioned but, as I see it, it is hard to see how an acceptable probability of failure of either traction control or electrically connected throttle control can be guaranteed on high mileage road bikes with no mechanical connection between the twist grip and the throttle, when they may not have had particularly good care or their fair share of maintenance.

I also cannot see how some electronic safety systems can easily be tested at regular intervals to prove they are still working. Upright traction control and ABS systems could be easily checked by the rider occasionally with a burst of throttle or a quick squeeze of the brake lever on a suitable road surface. The electronics could presumably be checked with special test equipment by a dealer but there are a lot more parts, both mechanical and electrical, in the total system besides the electronic controller. Cornering ABS or traction control at high lean angles would seem to be impossible to check without the somewhat risky idea of trying to provoke a slide to see if the system will allow it.

I did hear of one system of electronic throttle control which had two throttle butterflies in each inlet tract. One was under electronic control which took into account both twist grip angle, traction control and the programming of the control system and controlled the engine power in all normal circumstances. The second, manually operated from the twist grip, provided an independent means of shutting the engine down. This seemed to me to be a good compromise similar to modern aero engine systems, which have an electronically controlled fuel control unit to control engine power and a totally independent fuel shut off valve to shut the engine down if control of engine power was lost for any reason.

Basic ABS systems are in a better state with respect to the probability of long-term reliability, because of long term use in cars and the fact that most failures will result in the system reverting to normal manual braking, but the introduction of refinements, such as covering the effects of lean angle introduces more complication and potentially more ways in which the system could fail.

Modern aircraft rely totally on electronics for control of engines and flight controls and for the flight deck instruments which provide the feedback to the pilot. The systems and software are designed with the intention that all possible anticipated types of failure of the electronic systems are manageable and do not result in **Hazards** to the aircraft. It is interesting to note that a large proportion of defects requiring unscheduled maintenance in these systems usually relate to the various sensors, which provide feedback information to the control system on position of a movable control surface, or on temperatures, pressures, speeds, engine operating conditions, etc., rather than failures of the electronic control units. The same probably applies on a motorcycle, where wheel speed, throttle position, engine air flow and engine speed sensors, etc. and their associated wiring are essential to allow the control system to function and are all more exposed to weather conditions, corrosion, and general abuse than would be the case on an aircraft and are likely to get much less care and attention.

Aircraft sensors related to the control system are normally duplicated to cover the possibility of sensor failure, but this introduces another question. If two sensors at the same position give different signals, how does the control system know which one is correct? One way of countering this problem is to include a mathematical model of the normal operation of the engine or aircraft system in the control system memory. This will include the relationship between all the main sensor outputs over the whole range of engine or aircraft operating conditions, for example, if two engine rotational speed sensors at the same position in an engine show different speeds, the mathematical model could determine which is correct by comparison with the predicted engine speed for the same altitude, forward speed, throttle angle, compressor delivery pressure, exhaust gas temperature, etc. If all these other sensors are indicating as predicted it is not possible for the engine speed to be much different from that predicted by the model so the faulty sensor can be identified and automatically reported for maintenance.

I am afraid that I have wandered a long way off the subject again, but it all illustrates the basic fact that electronic systems, which may result in **Hazardous** conditions if they do not function as anticipated 100% of the time, should not be attempted without much thought as to the possible consequences of failure.

The following table, although by no means complete, provides some idea of the sort of things to look for if using any sort of electronic control in positions where **Hazardous failures** could occur.

Electronic Rider Aids		
Type of failure & possible effects	**Possible causes**	**Typical justifications for a satisfactory low probability of failure**
Engine power or braking not as required by manual controls or system programming. Possibly **Hazardous** depending on the type of deviation.	Deterioration of sensors or wiring due to damage or corrosion. Overheating of electronic units Failure of electrical power supply to systems or failure of the earth connections from the electronic units	Is the system exactly as supplied and tested by a major manufacturer, including all wiring and sensors? If not, suggest considering the possible effects of failure of the altered parts, including lengthened wires or similar small changes which do not affect operation but could affect the integrity of the system. Are the electronic units fitted in positions where there is sufficient cooling and well away from all hot parts? If not, suggest moving them to a cooler position. Have you reviewed the power supply, including any fuse provision and the earth connections, to the electronic units to ensure their integrity and the effect of any fuse failure? If not, recheck the system and review the probability and effects of a fuse failure.
Failure within the electronic units. Either in the control system or sensors. *or* In the manual controls, such as the twist grip, which select the required power or brake operation or in the actuators, such as the throttle actuator, which put the control system decisions into operation.	Failure of electronic or mechanical components of the units.	In this case we have to put total reliance on the system manufacturers as there is nothing the special builder can do apart from: Making sure that all the units are from a trusted source and, if second hand, have a known history, e.g. they have never been overheated or exposed to incorrect power inputs or short circuited outputs or in other ways damaged. Checking to ensure that there is no external damage to any unit.

PART 4

The Future of Motorcycle Modification and Special Building?

Predicting the future is always largely guesswork but one thing is certain. For all new mechanical, electrical or structural projects, the basic principles of failure analysis will remain the best way to ensure that the products are safe to use.

That is the four-stage approach, firstly to anticipate all possible ways any part of the product could fail, however improbable, then to identify all those where the effects of failure could be **Hazardous** and see if any of them can be eliminated by a design improvement.

Finally, to show that measures are in place to ensure that the remaining potential **Hazards** will never happen in the life of the product.

(Never is impossible to prove, but sufficiently improbable would be good enough)

Other than the above we are in the realms of speculation. These thoughts are my own and are offered for the reader to either agree, or otherwise.

As I see it, the largest threat to continued enjoyment of an interesting and satisfying pastime (or business) is the trend towards increasing regulation. Most politicians seem to have nothing better to do than thinking of new ways of preventing people doing what they want to do, even when it harms nobody, and the results of new regulation usually have little beneficial effect on safety or anything else. Two things are certain, additional regulation always means more delay and increases costs and once introduced, any new regulation is nearly impossible to get reversed.

The best counter to increasing regulation is continued vigilance and to point out that the pastime supports a large industry of parts manufacturers and restoration specialists. It is also the best practical training for interested young engineers, covering everything from stress analysis to workshop practice, project management and decision making. In industry most significant decisions are made by senior people, understandably because of the large sums of money involved, and the young engineer's task is usually to provide information to help with the making of these decisions and to make things happen once decisions on what needs to be done have been made. With a project under your total control, it is your decision what needs to be done and how, and success is much more satisfying and educational for that reason.

*What form future modification projects will take is open to speculation, but restoration of existing machines and what you could call cosmetic modifications, improving the appearance of machines without changing the mechanical or electrical features, will continue to be one part of the story. Widespread availability of 3D printing of parts would introduce more possibilities for manufacture of complicated one-off parts which would be difficult or impossible by conventional methods, but printing of **Critical** parts will need to be approached with caution if the necessary guarantee of material quality is to be achieved.*

Innovation in the real engineering parts of the machine is the most interesting part of the job in my opinion, and the direction this will take is open to debate. Some say the future is electrical and they may be right but perhaps further into the future than many imagine.

Motorcycles are in a much worse position than cars with respect t adoption of battery electric propulsion as minimising weight is essential for handling. In a car, a large weight of batteries can be mounted low down while retaining good handling characteristics although all the forces on the structure, suspension, steering, tyres, and brakes will increase meaning even more extra strength and consequently weight is needed.

Most motorcycle manufacturers seem to be experimenting with battery only systems and in general seem to be concluding that although battery power can be satisfactory for small commuter bikes, it is not possible at present (or in the foreseeable future) to match the performance, range, and handling of larger sports, touring or adventure machines. The fact that very few people will voluntarily exchange their current bikes for heavier, less capable, more expensive new electric ones is certainly concentrating minds in the industry.

This brings us back to the possibility of an alternative liquid fuel which I believe will definitely become available in the next few years. Using the large amounts of intermittent renewable wind and solar electricity, processes already exist to produce hydrocarbon fuel from almost any type of organic waste material close to the source of the power during periods when power demand for other purposes is lower and conversion of this to a petrol equivalent is likely to follow.

The first commercially available fuel of this type is already in production on a small scale, by the UK company, Coryton Advanced Fuels Ltd. Known as "Sustain Classic" the fuel contains no ethanol and does not displace any food crops from agricultural land. The cost is currently higher than petrol but would undoubtably come down with large scale production.

There is also the likelihood that a synthetic fuel produced directly by combining carbon and hydrogen will become available in the not too distant future. Racing is leading the way, as it often does, with Formula 1 cars and Moto GP bikes, among others, using increasing proportions of carbon neutral synthetic fuel in recent and future years. Synthetic fuel differs from bio fuel in that it is produced by direct combination of carbon and hydrogen but again the key driver is the availability of sufficient renewable electricity.

The only problem with respect to widespread adoption of carbon neutral liquid fuel would seem to be the high up-front cost of scaling up the process and the lack of political backing. Many politicians still seem to believe in the unattainable electric dream and refuse to consider alternative strategies. The reason I believe the dream of a large proportion of vehicles on the roads being battery powered in the foreseeable future is unattainable is not the capabilities of the vehicles but the financial impossibility of providing enough convenient carbon neutral charging facilities. We will get to this point again in more detail in the section covering my "best guess" General Predictions for Road Transport.

Collection of the waste organic material to provide sufficient feed stock for fuel production, without using good agricultural land for fuel crops, could perhaps be by means of a national collection scheme linked to existing recycling systems and collecting both agricultural and domestic waste.

The other possible carbon neutral fuel is green hydrogen produced by electrolysis of sea water. I believe it is a non-starter for general road transport although it may be suitable for some types of use. The problem with hydrogen is that it needs to be stored at very high pressure and has a very small molecule making small leaks impossible to avoid in general use. These would be no problem in the open air but could quickly form an explosive mixture if parked inside. Potentially blowing up the garage is not a good attribute for a new bike.

Another growing trend these days is towards over-complication with little thought as to whether advantages of the feature introduced are outweighed by the potential disadvantages. Not too much of a problem on bikes to date, but keyless entry on cars is a perfect example. The sole advantage would seem to be that you don't need to put your hand in your pocket and get the key out, but the disadvantages are numerous, including increasing cost, reduced reliability, possible failures which could lock you out and making the car much easier to steal. (Perhaps manufacturers employ too many clever people who need to keep thinking of these extra complications to stay in work.) Hopefully common sense will prevail eventually, at least on bikes.

Future Projects

In the meantime, there are plenty of options for individual projects, but matching available power to available tyre adhesion, without providing excessive unusable power and then reducing it by the use of electronic rider aids, would seem to be the best way to move forward on road machines.

One thought which came to me and could be done with available parts and technology would be what you might call a performance hybrid. Suppose you had a relatively small petrol engine, say 40 horsepower, just capable of maintaining a sensible maximum speed with little reserve power for acceleration at that speed.

Then add an electric motor and a relatively small battery with just enough capacity for extra power for a few bursts of maximum acceleration.

The electric motor torque could then be regulated to provide, when engaged, just enough extra torque at any speed to cover the gap between the torque provided by the petrol engine and the available tyre adhesion using something like the existing traction control systems to control the acceleration assisting torque supplied by the electric motor.

You could perhaps have a second stage to twist grip operation bringing in the electrical assistance when more than the available petrol power was needed. At all times other than during maximum acceleration or running at the selected maximum speed, there would be spare petrol power to recharge the battery. Recharging could be at quite a slow rate on a road bike, using a relatively small generator, as maximum acceleration is not needed very frequently and almost never for more than about 10 seconds continuously.

With such a system it should be possible to achieve superbike levels of acceleration at any speed up to the selected maximum speed without all the costs of running a superbike and without the great weight and high cost of a large battery with enough capacity for continuous running. (If it could be run on a synthetic fuel it would also be carbon neutral.)

Many current cars are advertised as self-charging hybrids, although they have sufficient battery capacity to allow a considerable electric only range. They do, however, point the way towards the proposed performance hybrid bike.

Second thoughts on the performance hybrid idea, (usually the best). Why not use electric drive all the time and run an even smaller petrol engine running at constant speed driving a bigger generator providing enough electrical power to maintain say a steady 70 mph without battery assistance, (about 25 hp)? This would eliminate the gearbox, most of the transmission system and the problem of integrating the two drive trains, allowing more space for the other part of the system. It would also simplify the controls as the petrol generator would control itself at constant speed, once started, controlling its own power in line with the electrical demand. The twist grip would only then need to control the electric motor.

Again, car developments are pointing the way with the Nissan "E power" system among others operating in this way.

Off the subject again, back to the past, mention of acceleration boosters reminds me of an aircraft engine I worked on in the 1960s and 70s which was intended to run for short periods as an acceleration booster during take-off and to be shut down for the remainder of the flight. The engine had originally been developed as a vertical take-off engine intended to lift a military prototype aircraft vertically during take-off or landing with a separate "horizontal" engine for propulsion. That aircraft concept never made it to production but the engine, which had a better thrust to weight ratio than any other at the time, was subsequently used as a propulsion booster engine on the Trident 3B aircraft for use on take-off only. The Trident was a medium sized airline aircraft and had three engines in the original version. The 3B variant was a stretched version of the aircraft with a higher maximum take-off weight.

There is an aircraft certification requirement which requires that a multi-engine aircraft must be able to continue to take-off and climb, at maximum weight, if one engine fails during take-off.

The maximum thrust of any aircraft gas turbine engine is limited by the maximum permissible temperatures in the turbine section of the engine, the atmospheric air pressure, which controls the mass of air which can flow through the engine, and the inlet air temperature.

Lower atmospheric pressure and/or higher air temperature both reduce the maximum rate at which fuel can be burned in the engine without exceeding the maximum permissible turbine temperature.

It follows that the worst conditions for achieving the rated take-off thrust from an engine are the hottest day temperature at the highest altitude airport from which the aircraft will operate.

With the increased maximum take-off weight, the 3B aircraft could not meet this requirement with one of the three original engines inoperative and there was no easy way to increase the thrust of the original engines.

Two of the original engines plus the booster gave sufficient total thrust to allow the aircraft to comply with the requirement. This booster engine was controlled by a 3-position switch in the cockpit; effectively, start to idle, boost thrust and stop.

The booster engine would then be dead weight for the remainder of the flight, but was a much smaller proportion of the total weight of the aircraft than a battery, generator and motor would be on a bike.

There is very little totally new.

You could say, "Back to the future."

One possible future development of the performance hybrid idea, which occurred to me, is a detachable engine/generator pack which could be easily removed for short range commuting leaving the machine still usable as a very light electric bike, and easily refitted when required. Two bikes in one.

Sketch 35 shows the general idea.

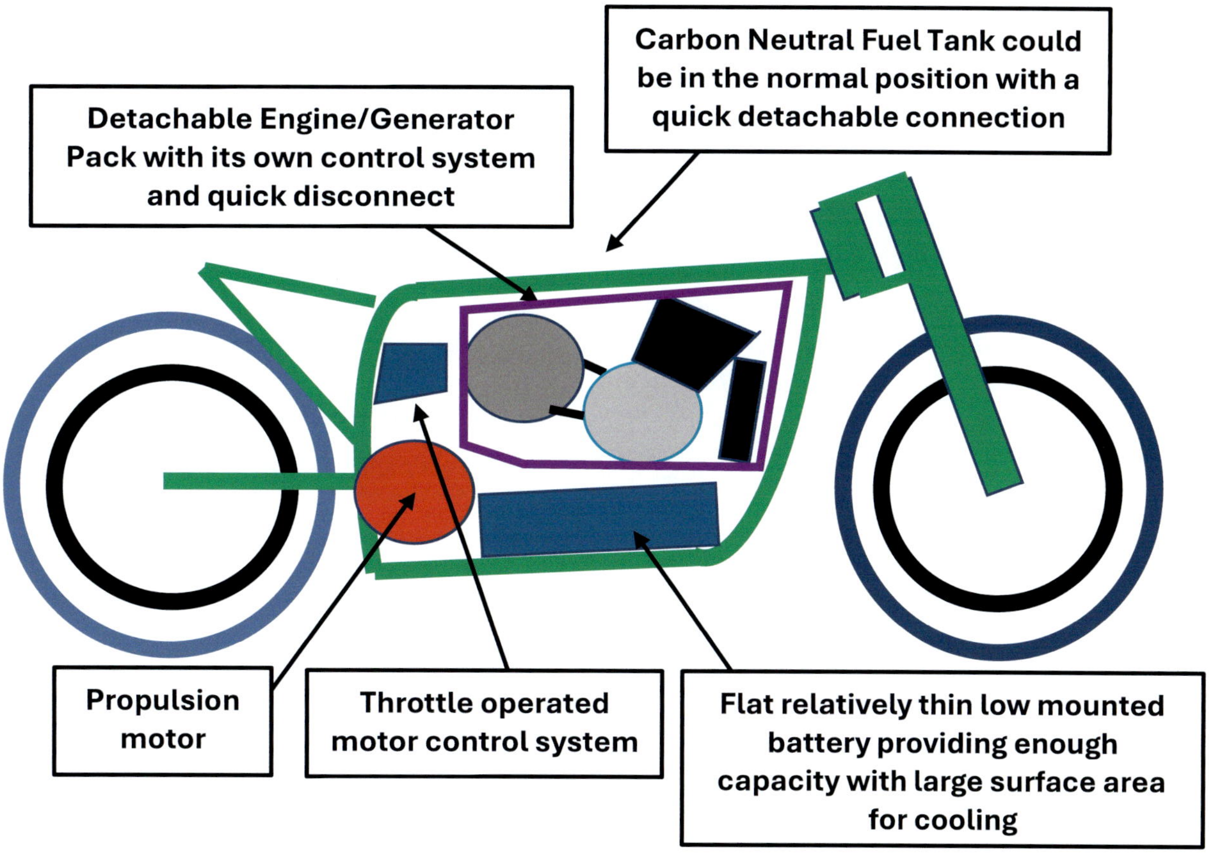

Sketch 35 – Possible future performance hybrid arrangement. Some sort of a simple lifting cradle would be needed for easy removal of the engine/generator pack, but it could also double as a stand to hold the pack when not in use.

The big difficulty with the performance hybrid suggestion, as with all systems which include a traction battery, is battery weight but compared with the weight reduction resulting from the smaller engine and gearbox, but with a relatively small battery it could be possible.

Another difficulty would be squeezing it all into a conventionally sized frame and ensuring that the battery was protected against accident damage or a rapid total discharge, but with the above suggested layout, it should be possible with a bit of ingenuity.

This brings us back to failure analysis. If you think about it, a traction battery stores as much energy as the equivalent quantity of petrol but this energy could potentially be released much more rapidly that the petrol could burn. The rate of combustion of leaking petrol is limited by the available air supply and the need to atomise or evaporate the petrol and mix with the air before it will burn. A satisfactory failure analysis of a battery powered bike would require systems to prevent very rapid energy release from the battery in any foreseeable failure or accidental damage case.

You could, with some justification, think of it as riding round sitting on a bomb, and trying to design precautions to prevent it ever exploding.

Battery fires are also not unknown with a number of incidences on cars, at least one on a double decker bus while in service, and numerous electric assisted pedal bike fires although most, but not all, occur as a result of overcharging. Whatever the reason, they are usually very intense and are almost impossible to extinguish before all the battery power is discharged.

Another potential ***Hazard*** *with any sort of electrical propulsion is the high voltage necessary to provide the necessary power at the rear wheel without very high electrical currents. Lower voltage and higher current arrangement would necessitate a very large heavy motor to produce the same power output and would waste more energy in heat. (If you kept the* *maximum motor current to say 100 amps, 50 hp would require more than 373 volts without taking into account heat losses and inefficiency, say 400 volts in total.)*

Insulation of this high voltage to avoid contact with the rider in any foreseeable failure or accident case would also be needed for a satisfactory failure analysis. It would seem that both positive and negative motor connections need to be insulated from the frame to avoid the whole bike and rider being charged to a high voltage insulated from earth by the tyres, which could result in something of a shock to the rider when he first put a foot on the ground.

Aircraft can build up a considerable undetected charge of static electricity in flight, with no ill effects on the passengers, but this needs to be discharged to earth after landing before anyone is allowed to leave the aircraft. You can see that the same problem could easily occur on a bike or car if the whole thing became charged to battery voltage undetected.

I have seen a press report of various electric supercar projects where they are running 800-volt systems. ***Lethal?***

Perhaps the eventual solution to this one will be special tyres with embedded filaments of electrically conducting material, but for now if there is any concern, a trailing earth wire in occasional contact with the road surface may not be a bad idea.

Aerodynamics

Looking further into the future, the question of aerodynamic assistance during acceleration, braking and cornering is bound to arise in racing although it is unlikely to be of much real value at normal road-going speeds. Ground effects to increase downforce on the tyres, as used on racing cars is not practical on a leaning motorcycle but some form of wing arrangement could be possible to provide downforce when upright.

Some race bikes, notably the BMW superbike and some Moto GP bikes are already incorporating small, fixed aerofoils to provide front downforce reducing the effect of weight transfer from front to rear as a result of aerodynamic drag. You can imagine that it would be difficult to get the bike to turn at very high speed if the front wheel was very lightly loaded. A fixed aerofoil works out very well in these circumstances as both drag and downforce rise with the square of the speed so a fixed wing angle can be the optimum at all speeds. BMW use two aerofoils per side, one above the other, a good way of reducing overall width.

Beyond this, the wing incidence angle would have to be capable of being changed during running to suit the different condition of straight line acceleration, constant high speed, braking and lean during cornering.

With a strong imagination you could think of four stub wings, one on each side at the front and back of the bike, all with individually adjustable incidence angles under electronic control. Then downforce and reduced weight transfer during acceleration and braking when required would be easy to arrange.

Perhaps the downforce could be arranged to act directly on the wheel spindles through a linkage to avoid compressing the suspension, although this may be a step too far. Direct connection of wings to the un-sprung parts of the suspension was used on several Grand Prix cars a long time ago but subjected the wings and linkages to high vibration and shock loads due to suspension movement, resulting in high fatigue loads. Movable aerodynamic devices were then banned in car racing after several failures, which could leave the car completely out of control if they occurred at the wrong moment and resulted in at least one fatal accident.

Sketch 36 below shows how aerodynamic downforce could be used on an upright bike.

Acceleration **_High front down force to keep the front wheel down with weight transfer to rear. Some rear down force to assist traction._**	***High speed*** **_Some front down force to counteract aerodynamic drag induced weight transfer to rear._**	***Braking*** **_High rear down force to counteract weight transfer to front, allowing effective use of both front and rear brakes._**

Sketch 36 – Possible stub wing generated downforce. Aerofoils shown red.

It is hard to see how any extra useful vertical downforce at the tyre contact point could be generated during cornering because fixed aerofoils would lean with the bike.

Even further beyond this first use, perhaps the way forward would be to use the aerodynamic forces on the stub wings to reduce lean angle during high-speed cornering while generating minimal vertical or lateral forces. Again, it is unlikely that much lean reduction could be generated at normal road speeds but at racing speeds, it could be very effective. (A very approximate calculation, based on relatively small stub wing areas and an overall width not much more than some monster cruisers, shows that some usable positive or negative lift forces could be generated at about 70 mph and, as lift increases with the square of the speed, all other things being equal, there would be more than enough available at 100 mph and above.)

From a practical design point of view, the maximum lean angle achievable without too much grounding, with any reasonably sensible footrest height and width, would seem to be as serious a limit to cornering speed in racing as tyre adhesion. If an aerodynamically generated torque in the same direction as the gravity effect of lean could be generated, it could replace some of the lean requirement at speed and allow development of better tyre adhesion to be exploited without necessitating totally impractical lean angles. Sketch 37 shows the general idea.

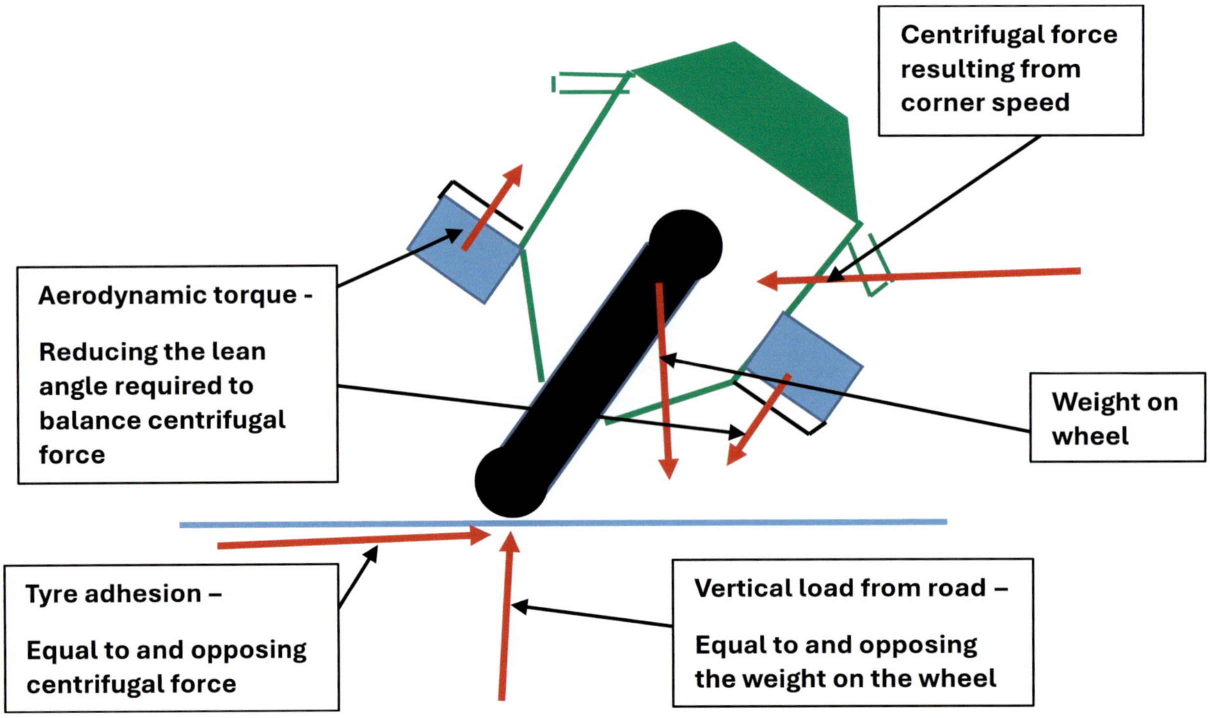

Sketch 37 – The suggested aerodynamic lean reduction system.

Controlling the aerodynamic torque would require some ingenuity as, unlike acceleration and braking, where operation of the throttle or brake lever could trigger the optimum levels of downforce, cornering requires anticipation. Perhaps it could react to the initiation of normal lean relative to a gyroscopically defined vertical. (It could be called an ALR system (aerodynamic lean reduction) as everything seems to need to have a mysterious abbreviation these days.)

With a bit of practice, such a system could feel quite natural and intuitive. After all, we quickly adapt to the differences in lean angle, at the same cornering speed, needed with wide or narrow tyres or on the same bike in cross winds.

*If some failure in the system selected lean reduction when travelling in a straight line at normal road speeds, it would feel like an unexpected sudden cross wind gust and, with the built-in limit of say not more than 10° lean reduction, could be countered in much the same way as a wind gust. At racing speeds incorrect operation would be much more **Hazardous** as the aerodynamic forces would be higher and the time available for correction much less.*

Again, programmed computer control for the aerofoil incidence angles would be needed to optimise levels of aerodynamic torque to suit the forward speed and the cornering situation and to limit the effectiveness of ALR to the suggested 10° maximum.

The first snag with this sort of system is the likelihood of movable aerodynamic devices being immediately banned in racing, if they showed some promise, as they were in Grand Prix cars many years ago. That ban has resulted in cars with large aerodynamic downforce at all times necessitating very hard suspension to support the downforce and creating a lot of aerodynamic drag. The hard suspension, and the need to control under floor clearance to maintain downforce, results in cars which are unable to cope with even moderately rough road surfaces.

A demonstration lap by a Formula 1 car at the Isle of Man TT was suggested about 30 years ago because the cars were much faster, on the very smooth Grand Prix circuits, than the bikes, mainly because of aerodynamic downforce. The idea was rejected almost immediately, as the car would have been totally incapable of dealing with the IOM road surfaces which were, and still are in some places, not exactly smooth, level, and seamless. In its place, a modified road car, a 3.5 litre Rover Vitesse, did the lap and achieved a 100 mph average speed. The driver was BL works rally driver Tony Pond and the lap was a great achievement in that car, although it was considerably slower than the bikes. The car would have had a much lower top speed and less acceleration and would have been further handicapped by its width making all the corners tighter and producing more corners in places which could be straight lined on a bike.

It would be interesting to see what a modern supercar, such as a McLaren, could do on the TT circuit. It would certainly have the speed and acceleration to match the bikes and would benefit from some level of aerodynamic downforce, but its width on the relatively narrow twisty parts of the course and the driver's lack of detailed course knowledge compared with TT regulars may be the limiting factor. Perhaps someone will organise a demonstration lap one day.

More recently, the ban has been partially relaxed in Formula 1 to allow aerodynamic drag reduction in some circumstances for straight line overtaking but is still in force for all other running.

The second snag is that, if the front end lifted at high speed as a result of road conditions, not related to aerodynamics, some means of avoiding increasing front end aerodynamic lift would be necessary, as this would make getting it down again much more difficult. This would be easy to achieve if aerodynamic lean reduction was not included. Upwards lift would never be needed, and the stub wings could be arranged to fold upwards and become ineffective if any upward force was applied to them.

If "ALR" was included requiring both up and down lift, it becomes much more difficult.

The third snag is back to failure analysis. Any incorrect wing angle or delay in changing the angle could be ***Hazardous*** *at racing speeds and not too good on the road. You would be relying totally on the control system and the powered operation of the wing actuators to function correctly 100% of the time. I am not sure that such systems are reliable enough for this at present but who knows in the future. They say we are going to be overrun with driverless cars which will need similar foolproof, and everything else proof, systems if they are to become a reality.*

Wandering further off the subject again, (hopefully for the last time) the mention of driverless cars prompts the thought:

"I hope somebody is doing an aircraft type failure analysis of these driverless machines."

That task would seem to be something of a challenge with the thing needing to remain safe while carrying any combination of sensor, system or mechanical faults which could not be positively proved to have an acceptably low probability of occurrence, and while operating in an unregulated road-going environment without any human monitoring. I guess that there are more sudden unanticipated events, such as other drivers doing the wrong thing or wayward pedestrians or animals, in a single road journey than the average aircraft sees in its whole life. In this environment, the machine would have to look after itself between periodic manual system checks which would definitely need to be compulsory, frequent and signed off by a responsible person. I don't think I would be a volunteer to sign off such checks, knowing that a single mistake during a whole career, with no human drivers to notice and report any problems, could cause mayhem and lead to a potential manslaughter charge.

Fortunately, I don't think there is any prospect of riderless bikes in the future. No doubt it could be done but what would be the point?

Predictions for Future Motorcycle Development

Finally, I will have a go at predicting what the new bikes of the 2030s will be. My best guess is that almost all small, short-range bikes will be battery powered but those needing higher power and range will still have an internal combustion engine which will run on a carbon neutral liquid fuel. There will still be large numbers of older bikes on the road, adapted for the available fuel, probably outnumbering the new ones by a large margin for many years.

Superbikes and all other "cutting edge" machines will have electric transmission with no direct drive from the engine to the wheels. Their engines will be somewhat smaller than current superbikes and will drive a generator continuously charging a relatively small traction battery and would be automatically controlled to match the electrical demand. The battery would provide additional power when required for acceleration or very high speed. The battery could be mounted low down in the current gearbox position and drive an electric motor at the swinging arm pivot. The smaller engine could be turbocharged to improve efficiency and save weight as a turbo's slow "throttle" response would not be an issue.

With no gearbox or clutch there will be no need for foot controls and the throttle, and both brake levers could be on the handlebars. The electronic controls systems will be further refined to limit maximum rear wheel torque and front and rear braking to match the available tyre adhesion both when upright and when leaned.

Moveable aerodynamic features are unlikely on road bikes but there will probably be fixed front end downforce to cancel out the front end lift due to aerodynamic drag and keep the proportion of weight on each wheel constant at all constant speeds. For racing speeds an additional movable rear aerofoil would be a possibility to apply rear downforce during braking to maintain rear wheel in heavy contact with the road during periods of braking weight transfer to the front.

Time will tell whether I am right but if things did develop as suggested it would be the biggest step change in motorcycle design since the 1920s.when the present day arrangement of engine, clutch, gearbox and chain or shaft drive to the rear wheel was generally adopted.

I will be too old to ride then but hope to be still here to see how things develop.

General Predictions for All Future Road Transport

A fairly safe prediction is that the range extender type of hybrid using carbon neutral liquid fuel will be the preferred vehicle for general use in the foreseeable future with pure battery electric propulsion restricted to use where sufficient charging facilities are available.

Although current government policy in the UK appears to favour almost 100% battery electric systems, I believe that is based on wishful thinking and the influence of pressure groups rather than reasoned engineering and economic judgement.

My reason for this conclusion is not based on the capabilities of battery electric propulsion which will no doubt improve over time and could match internal combustion with liquid fuel eventually.

It is based on the sheer financial impossibility of providing sufficient battery charging infrastructure in every area of the country. Whatever rapid charging capability is achieved with the batteries, the charger still has to feed the necessary number of kW within the reduced charging time.

As an example, a 5 minute charge of 50 kW hrs (or 4 simultaneous 20 minute charges) would require more than a 12 X 50 = 600 kW from the local power supply, equivalent to the peak power demand of about 60 normal houses.

Bearing in mind that the power transmission system must cope with peak demand from both increased requirements for existing users and the new charging systems simultaneously, rather than the 24 hour average often quoted, it is hard to imagine how sufficient electrical power could be transmitted from remote renewable generators to where it is needed to allow a sufficient number of chargers in the right places in all areas of the country even if the required green electricity was available at source.

Perhaps much further into the future, local power generation using small modular nuclear reactors or even fusion power may be possible but that is a very long way away.

Last Thoughts

Enough of the speculation, my interests are much more modest: I intend to continue the development of the trikes with ideas for various improvements.

Other than that, my next project will be to produce an ultra-light, ultra-simple two-wheeler, with nothing on it which it does not need to run, in order to keep me on the road for another few years. I still have the horizontal Triumph (photos 14 and 15) stored in fairly tired condition as the basis for this machine. A restoration and modernisation of this bike with a modern electrical system and disc brakes, and a serious programme of weight reduction would be a good basis.

One thing I have found during all the years of working on aero engines and bikes is that the key to weight reduction on an existing machine is often not a single big change but is by looking at every individual part and, if it must be retained, reducing the weight of each, however small the possible reductions may be. They all add up.

The mention of weight reduction has prompted another memory from the past which illustrates the point well. At one time I worked on a new aero engine type which had been certified for service use, but the first engines manufactured for production aircraft were somewhat over the maximum weight specified in the contract with the aircraft manufacturer. This mattered because it meant paying substantial penalties for every over weight engine delivered. The chief engineer at the time offered to pay additional manufacturing costs of up to £100 per engine set for every 1 lb weight saved and the necessary weight reduction was soon found, although a large number of modified parts had to be re-approved by the airworthiness authorities. From that time on, all production engines were under the contract maximum weight which benefited both the engine manufacturer and the aircraft.

Finally, I hope those who have persevered and reached the end of this part of the book may be inspired to attempt a few modifications to their machines. The sense of achievement, when even a very small modification works well, is such that it is likely to inspire more modifications becoming more radical as confidence grows.

Part 5, which follows, has been added to the second edition of the book in order to pass on experience gained during conversion of two-wheeled bikes into three-wheelers. Hopefully it will help anyone considering a similar conversion.

PART 5

Adding an Extra Wheel

Photos 32 and 33 – A transformation.
Base bike to Mk 2 three-wheeler which later became the Mk 2.5 with improved nose section.

Introduction

Adding a Third Wheel

As I have spent most of the last few years on motorcycle based three-wheeled projects, the second edition of the book provides an opportunity to pass on a few of the things I have learned, sometimes the hard way, about the peculiarities of 3-wheeled machines in the hope that they may be useful to anyone considering a similar project.

Although three-wheeled problems are not directly within the scope of the original purpose of the book, failure analysis is an essential part of ensuring safety with this type of project where there are very few similar machines to use for guidance.

As with the previous sections, I have avoided much engineering theory and tried to produce a practical man's guide to avoiding the many pitfalls in a project of this kind.

I have also continued the previous system of using italic type to separate out my own experiences from the factual material.

The first requirement of a 3-wheeled project is to decide what type to build. The easiest is obviously the sidecar with a variety of parts and vast experience readily available, but sidecars with their asymmetric handling and other peculiarities are not for everyone and certainly not for me after some small experience many years ago.

This leaves trikes with two at the rear or two at the front as the options and a choice between the motorcycle type, steered with handlebars, or the 3-wheeled car type, steered with a steering wheel.

I chose the handlebar steered two at the front type, often called a reversed trike, as I wanted to retain as much of the motorcycle riding experience as possible with the prospect of better handling and, in my opinion, safety in difficult conditions compared with a conventional trike with two at the rear and a motorcycle front end. With a normal trike excessive understeer, particularly in slippery conditions, is likely to be a problem as the weight distribution is always going to be biased towards the rear.

The reversed trike is definitely the most difficult to make of the various types and the experience gained is the basis of this part of the book. Much of it would also be applicable to a normal trike.

What Type of Bike to Convert

In most cases this decision will have already been made as the bike will be your own two-wheeler. This was true my case as I had used both bikes previously in two-wheeled form. If there is a choice, it is better to go for something with good low speed torque than one which produces a lot of power at very high revs but is a bit lacking at low engine speeds.

Given a free choice, I would go for a touring type V twin of about 600-750 cc, i.e. the equivalent of a larger more powerful version of the 400 cc Honda Bros which I used in my first attempt. But everybody has their own ideas.

1 – Basic Principles and Overall Dimensions.

Before starting on any project, it is always better to settle an outline of what you are trying to achieve. A manufacturer would spend thousands of hours designing every detail before making anything, but to some extent an individual project designs itself during the build in that you tend to start with the concept and basic structure and the design evolves during the build, as the various parts are fitted into wherever seems best.

For a reversed trike, I believe the following things are essential and should be settled before starting.

WHEELBASE AND WEIGHT DISTRIBUTION

For reasons of weight distribution between front and rear it is best to keep the wheelbase as close as possible to that of the original bike, although a small increase to make room for the new front structure is difficult to avoid. A forward biased weight distribution is essential for both cornering and aerodynamic stability and a centre of gravity a third of the wheelbase from the front with the rider on-board giving equal weight on each of the three wheels seems optimum.

Most bikes have a centre of gravity close to the mid-point of the wheelbase and a good approximation for the position of the rider weight is 2/3 on the back and 1/3 on the front, so you are starting with a rear biased weight distribution.

The extra weight of the new front structure and extra wheel helps to move the C of G forward but moving heavy items like the battery to the front is generally beneficial.

I managed to keep the increase in wheelbase down to less than 2 inches in both my attempts with close to equal weight on each wheel with a 150 lb rider on board, so it can be done.

Sketch 38 shows the sort of weight distribution to aim for and getting close to this is fundamental to achieving good stability and handling for the finished machine.

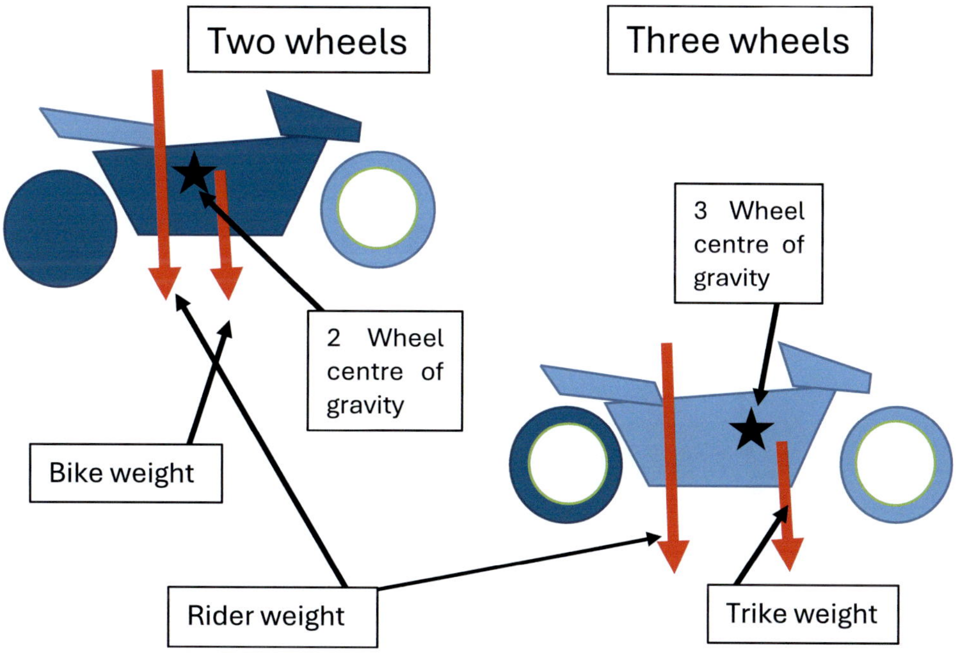

Sketch 38 – Comparison of best centre of gravity position for machine plus rider.

FRONT TRACK WIDTH

Front track width between the centre of the road contact area of the front tyres is another early project decision to be made. There is, as far as I can see, no absolute best ratio between wheelbase and track width, but there are limits to stay within.

The most important is the minimum track width necessary to avoid the danger of overturning on corners or if the machine gets a bit out of line in slippery conditions and suddenly finds grip. The unfortunate thing about three wheels is that the effective width in these circumstances is less than the front track width. The machine rolls about the lines between the front and rear wheels on the outside of a corner so the effective width is the distance between these two lines at the centre of gravity position. If the centre of gravity is at the desirable 2/3 wheelbase from the back, the effective width will be 2/3 of the front track and you should aim to ensure that half of this effective width is preferably more than, or at worst, not much less than the height of the centre of gravity.

To put some numbers to it, if the reversed trike, including rider, had a C of G height of around 16 inches (40.6cm) the front track width should not be much less than around 3/2 x (2 x 16) = 48 inches (107cm). If you want to go much below this width you really should be thinking of a leaning trike design which is a whole lot more difficult, (or not attempting much fast cornering).

In general, a large track width and a low centre of gravity will provide security against overturning but going too wide relative to the wheelbase will make the trike difficult to control on rough surfaces as single wheel bumps will tend to deflect it towards the bump side.

There is also a limit on overall width of 1.3 metres (51.17 inches) imposed by UK IVA construction and use rules if you want to retain the single motorcycle headlight.

In my opinion a track width of 75 to 80% of the wheelbase is about the optimum providing a good compromise between all the various factors.

Both my machines are within this 75/80% rule and have shown no problems related to this aspect.

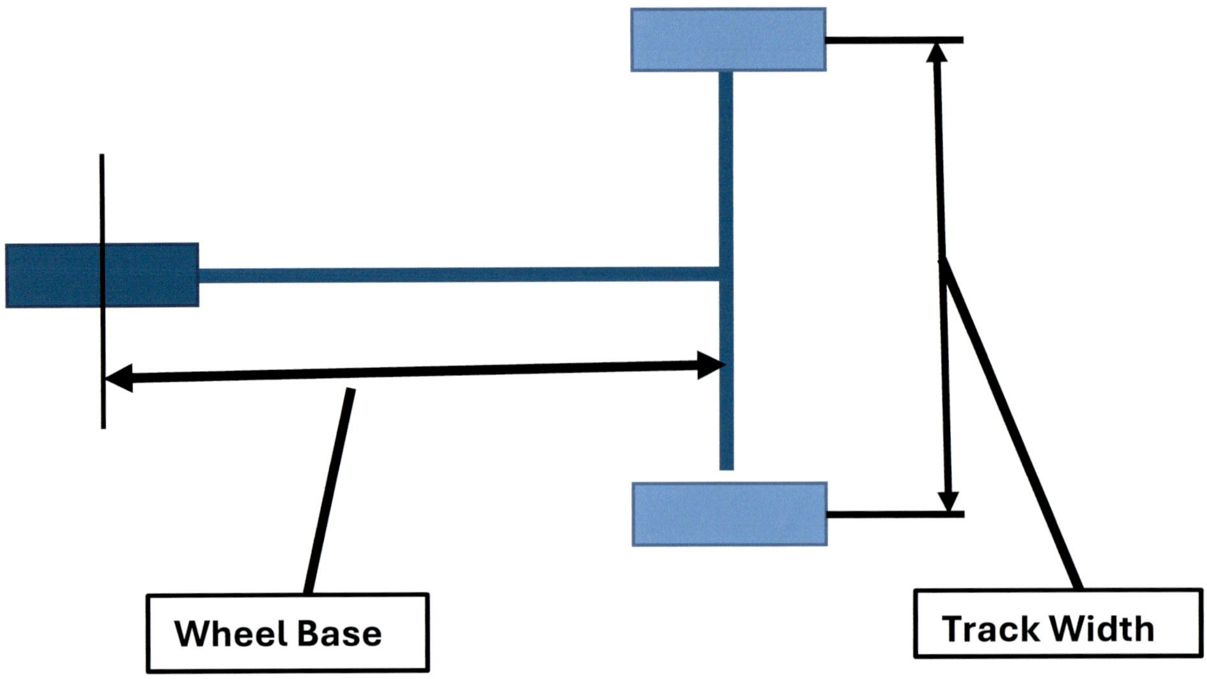

Sketch 39 – Wheelbase and track width.

To Lean, or Not to Lean

As mentioned above, there is always the possibility of arranging for the complete machine to lean like a two-wheeler on corners, which would certainly be closer to the motorcycle riding experience.

It is, however, not easy to achieve and the perceived benefits in increased cornering power are really very limited, apart from with machines with a very small front track width. (A 30° lean would only move the centre of gravity over by 5.3 inches in the above example.)

In my opinion, provision of leaning is not worth the extra complication on most machines and also, by necessity, must include many more ***Critical*** *components and joints in the front suspension. It could become a Failure Analysis nightmare, although there are a number of very narrow track leaning three wheelers on the market to use for guidance.*

2 – The New Front Structure

CONNECTING TO THE BIKE

The main structural components of the new front end are a new front sub-frame to locate and support the front suspension and all other new front end parts and the means of connecting the sub-frame to the bike.

It is best, in my opinion, to build the front sub-frame as an almost complete unit separately before connecting it to the bike as this gives less weight and better access for ease of working, *(you can turn it over rather than working underneath when you need to).*

First, however, the points on the bike where it is to be connected need to be settled as the sub-frame can then be designed to suit.

On most bikes, the strong points most suitable for attaching the front end are under the steering head, the lowest front engine mount position and somewhere close to the rear suspension swinging arm pivot.

It is not a good idea to attach the sub-frame by welding for three reasons.

1) Guaranteed weld quality would be needed, as failure could be **Hazardous**
2) Retention of an unmodified frame would allow easy re-conversion to two wheels if needed in the future.
3) An unmodified frame is a definite plus point when it comes to UK re-registration as a light three-wheeler.

Bolted joints are therefore best at all connection positions.

I have used a connecting structure comprising of 11 or 12 tubular struts which has proved satisfactory to date. Sketch 40 show the principle used.

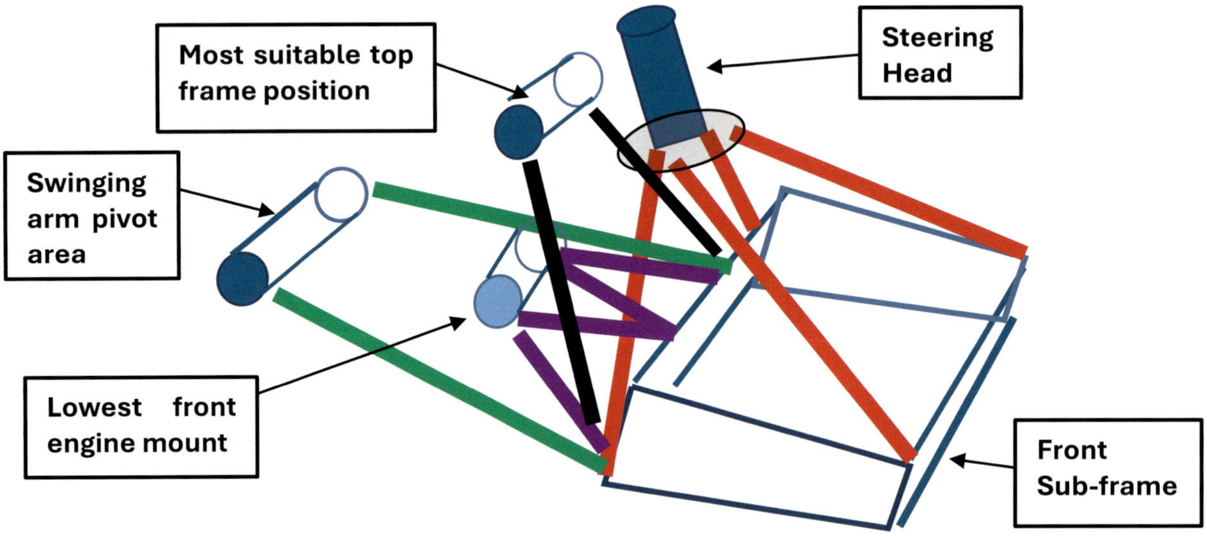

Sketch 40 – Connecting structure principles as used on my own projects.

This results in what is known as a redundant structure in which disconnection of any one strut does not cause the structure to collapse.

It does mean that the share of the loads carried by each individual strut is difficult to access and the simple way to ensure sufficient strength is to think of it as four separate structures, each with its own purpose. Then any load sharing will tend to reduce loads in the primary structure.

- The red struts carry the weight of the bike supported from the steering head in the same way as the weight is carried on the base 2-wheeler.
- The green struts locate the new front sub-frame in its fore and aft position and transmit drive and braking loads.
- The triangulated purple struts locate the sub-frame in a transverse direction.
- The black struts keep the bike vertical relative the sub-frame.

Not the lightest way of connecting the sub-frame but definitely the easiest to make and the proportion of overall weight is very small.

The system has worked out very well on my projects with no problems in use and has the advantage that building in some adjustability in some of the strut attachments allows accurate alignment of the sub-frame.

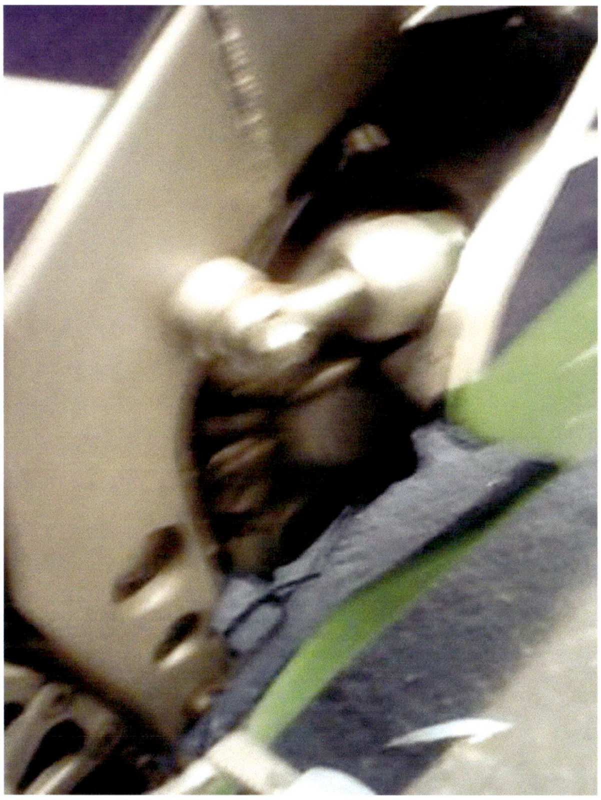

Photos 34 – Parts of the front connecting structure of my 2nd machine (the Mk 2).

Most of the struts are made from 22mm x 3mm thick stainless steel tube with the diameter of the long strut to the swinging arm pivot increased to 38mm diameter to avoid possible buckling problems. The strut ends are all attached by simple bolting. Photo 34 shows typical parts of the Mk 2.

The red struts are attached at the steering head by means of a long bolt through the hollow steering stem. This obviously must not introduce any frictional resistance into the steering and the easiest way to achieve this is to use a small car or trailer wheel hub which gives you ready-made taper roller bearings and 4 studs to attach the four "red" struts. With a stepped sleeve to align the two internal diameters and the long through bolt, this provides a perfect almost frictionless top mounting. I used a BL Mini rear hub on the Mk 2, and it has proved far better than the large car ball joint used on the Mk 1. There are lighter 4-bolt hubs from trailers, etc. which would be even better. Sketch 41 shows the basic idea.

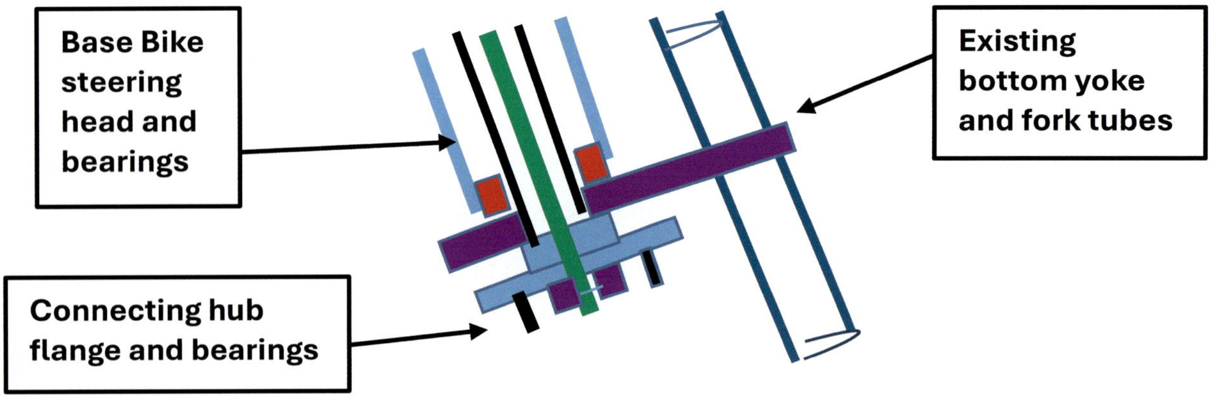

Sketch 41 – Connecting the front structure to the existing steering head with a 4-stud hub flange and bearings.

3 – Front Sub-Frame and Front Suspension

This is the most difficult part of the whole conversion to get right as there are two conflicting requirements affecting the suspension, and finding best compromise is not easy.

These are the need to minimise roll during cornering and the need for low spring rates, lower than you would at first assume, to accommodate small ripples in the road surface.

The suspension geometry therefore needs to be settled early and the remainder of the front end designed around it.

The easy way is to use a front sub-frame from a small car as the basis and modify it to suit.

I did this on the Mk 1, using a BL Metro front sub-frame complete with the suspension linkage and steering parts. The width was reduced by cutting it in half at the centre line and removing material to achieve the required track width, then welding the two halves together.

Motorcycle suspension units were fitted as it proved impossible to reduce the spring rate of the existing hydro gas units sufficiently.

This is reasonably successful, if somewhat heavier than necessary, but tends to roll onto the bump-stops on corners, even at moderate cornering speeds, which could make an unexpected mid corner bump a challenge.

The reason for the excessive cornering roll is that the front roll centre of the car type double wishbone suspension is much lower than the centre of gravity of the bike plus rider and the rear roll centre, with the single wheel, is at ground level. Sketch 41 illustrates the problem.

The Mk 2 front sub-frame is shown in Photo 3 in Part 2 Chapter 3 and was based on two aluminium plates which sandwiched the suspension mounting points between them and provided attachments for the connecting struts. Photo 35 shows a top view of the basic sub-frame.

Photo 35 – Top view of the Mk 2 bare front sub-frame.

3.1 – ROLL CENTRE

The roll centre is the line about which the machine will roll when it leans as a result of cornering side force. When a side force on the machine results from cornering, it effectively acts at the centre of gravity and if this is above the roll centre, the machine will lean outwards. This effect does not have a great effect on cornering unless it hits the suspension bump stops or lifts the inside wheel, but does not feel right, especially to a motorcyclist used to leaning the other way. On a reversed trike, the roll centre at the back is at the single rear tyre ground contact point, much lower than on a four-wheeler, so the roll centre at the front needs to be as high as possible so that the overall roll centre is as close as possible to the centre of gravity. Sketch 42 shows how to find the roll centre with car type double wishbone suspension and shows that it must always be very low and will get lower as the suspension compresses. Not what we really need on a reversed trike even though it is the easiest to produce from car parts. *I used this system on the Mk 1, my first attempt, and it worked well apart from the tendency to corner on the outside front wheel bump stop if any sort of enthusiastic cornering was attempted.*

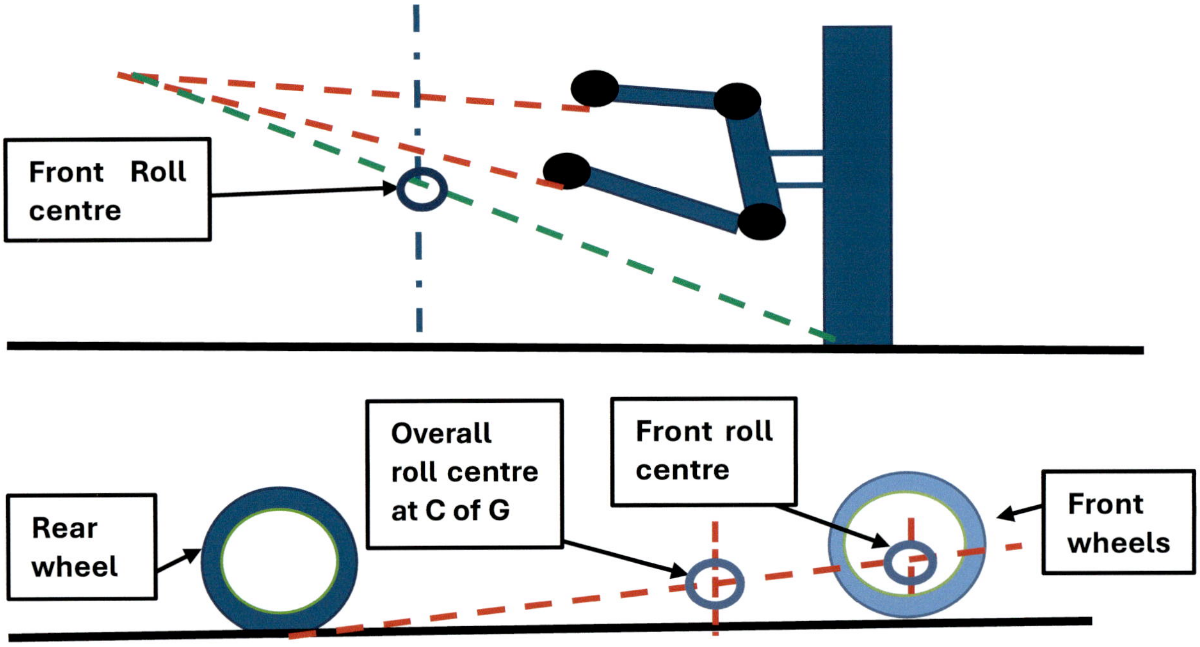

Sketch 42 – Roll centre with double wishbone suspension.

The only real solution to this difficulty is to use a front suspension system with a higher roll centre. Anti roll bars, connecting both front wheels with a torsion rod spring, as are often used on cars, have the effect of transferring some of the extra cornering load on the outside wheel suspension spring to the inside wheel spring, and thus reduce the overall roll angle, but they also affect suspension operation during a single wheel straight line bump, of which more later, so are not a total solution.

There are several ways of achieving a high front roll centre, one of which is the high pivot swinging half axle, as shown in Sketch 43. This system is not often used on cars as the wheel lean angle changes as the suspension compresses which does not suit flat tread car tyres.

The motorcycle tyres I have used on both my machines are not affected by a few degrees of wheel lean and the system I have used on the Mk 2, the second machine, approximates to the swinging half axle geometry while allowing the use of the same Metro swivel hub components as used on the first machine, Sketch 43.

Other possibilities include trailing links, McPherson struts, the vintage Morgan sliding pillars or even a lightweight beam axle which may seem old fashioned but provides a roll centre at wheel centre height and could work very well on a reversed trike if well-located fore and aft and transversely by an arrangement of links. The link angles could even be arranged to provide anti-dive braking, *(an interesting possibility for a Mk 3 if there is one).*

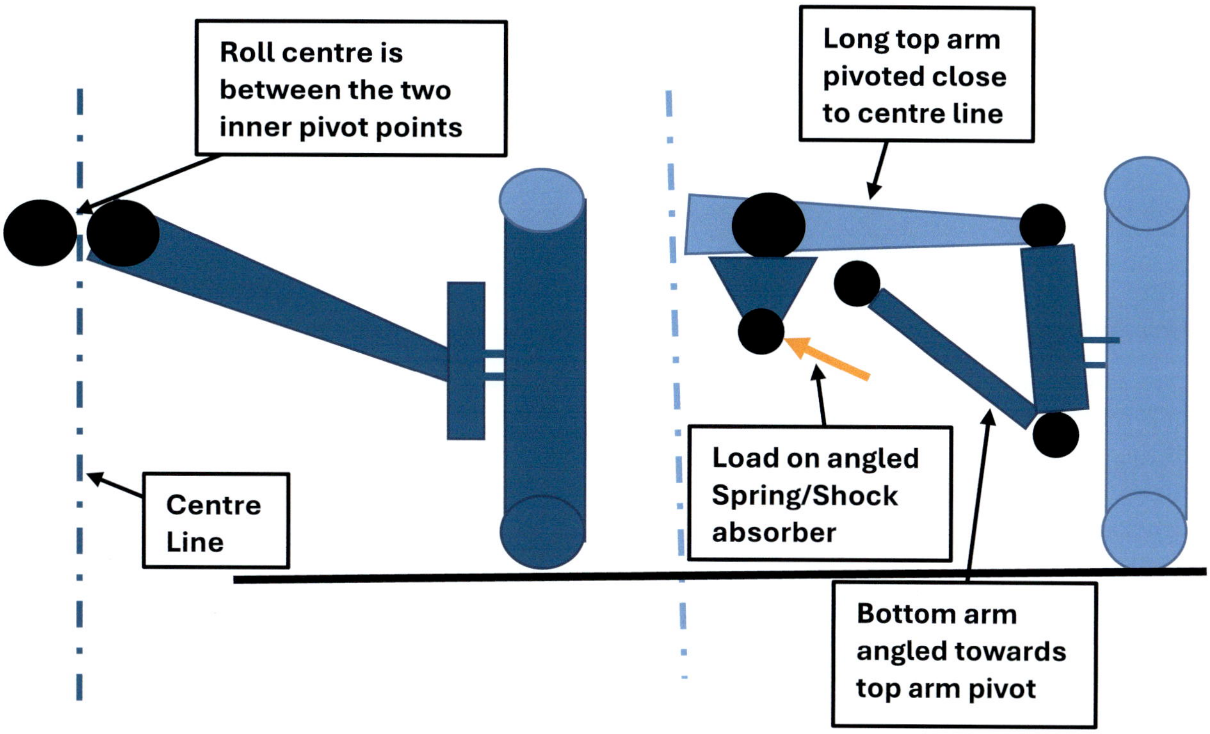

Sketch 43 – The Swinging Half Axle principle and my approximation to it on the Mk 2 machine (right).
The approximation provides a front roll centre 6 inches above the wheel centre and not far from the height of the centre of gravity of the whole machine. (Photo 3 in Part 2, Chapter 3 shows the general arrangement.)

Once details of the suspension system are settled the remainder of the front sub-frame forward of this point is generally lightly loaded, apart from providing strong points to attach any forward links from the suspension system, and can be designed to suit the various components and bodywork parts to be fitted.

3.2 – SPRINGS

The second aspect of front suspension is springs and shock absorbers and there are a variety of motorcycle units available which could be used, or specials can be made by companies such as Hagon-Shocks.

The main difficulties are settling the best spring rate at the wheel and the load in position applied at the spring location when static with the rider on board.

The spring load in position is the "corner weight" with the rider on board, less the weight of the wheel and those components which are also un-sprung and move with the wheel, and multiplied by the ratio of wheel movement to additional spring compression,

(At an early stage in the absence of measured weights a reasonable estimate based on my machines would be half the weight of the original bike, less its wheels, on each front wheel)

The best spring rate is the more difficult aspect of the front springs, and it needs to be much lower than you would at first assume.

On my machines the existing fork springs on the base bike provided a combined rate of around 100 lb/inch at the wheel but the same rate on the trike on each front wheel proved to be far too hard even though the weight on each wheel was similar to that of the original bike.

The reason for this is best seen by imagining a trike with the wheels directly fixed to the frame structure with no springs. Then a 1-inch single front wheel bump would lift that corner by 1 inch.

On a two-wheeled bike, that would lift the centre of gravity, at about mid length, by ½ inch, pivoting about the back wheel.

Similarly, on a 4-wheel vehicle the 1-inch bump would lift its centre of gravity on the centre line by approximately ½ inch, pivoting on a line through the wheel contact points on the other side.

On a trike, this pivot point is on a line between the front wheel on the other side and the back wheel and is much closer to the centre of gravity. So, the 1 inch single wheel bump will lift the centre of gravity much less than ½ inch but it will still feel like a 1 inch bump at handle bar, etc. That means that the momentary upward load needed to accelerate the wheel and frame corner upwards is lower.

Sketch 44, based on dimensions of my second machine, illustrates the effect with the centre of gravity at 1/3 wheelbase from the front.

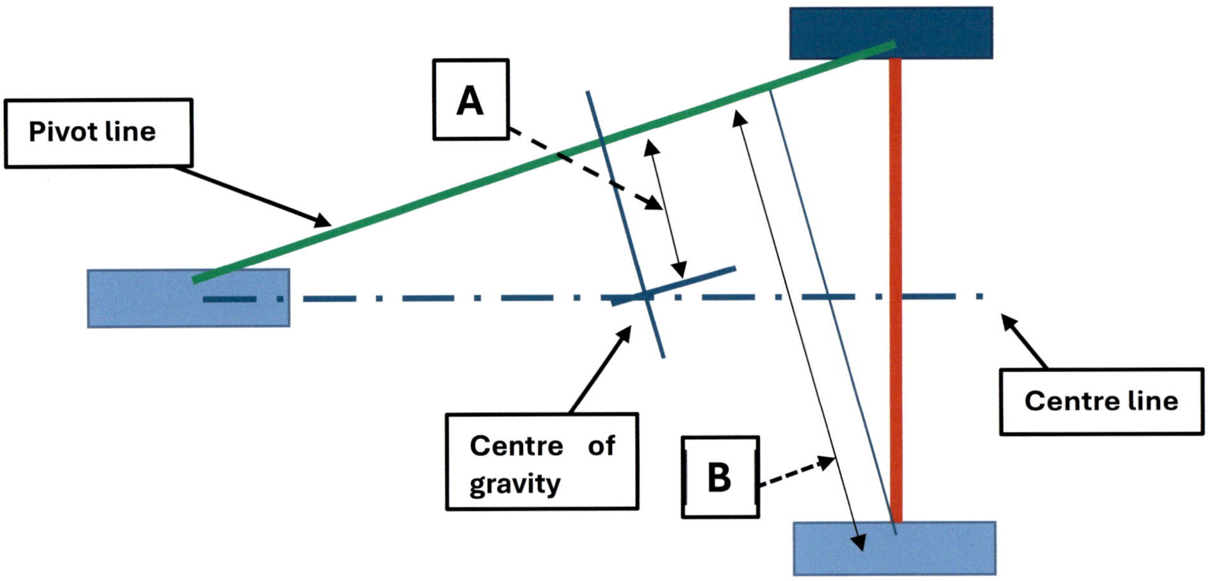

Sketch 44 – Basic reversed trike geometry. Dimension A is approximately 1/3 of dimension B.
On a two-wheeler with a central C of G, dimension A would be ½ of dimension B, although in that case the dimensions would be along the bike in place of across.

When we think about one front wheel riding over a 1-inch-high bump, with no suspension, the upward acceleration of that corner of the frame would be fixed by the forward speed and shape of the bump regardless of the number of wheels. Assuming equal weight on the wheel, the force applied to the frame by this upward acceleration would, however, be much lower on a reverse trike than on either an equivalent 2- or 4-wheeled machine because the trike will have a lower "Moment of Inertia" about its pivot line.

The "Moment of Inertia" is basically the sum of the weights of every component of the machine multiplied by the square of its individual distance from the pivot line. (You could think of it as the resistance to changes in rotation speed in the same way as mass provides the resistance to changes in linear speed.) The majority of the heavy parts will inevitably be close to the centre line, and therefore closer to the pivot line than on either 2 or 4 wheelers, so the upward force to accelerate a front corner will be much lower than on either comparable 2 or 4 wheelers.

As an example, a 1 lb weight placed over the bump side front wheel would have something like $3^2 = 9$ times the effect on the moment of inertia about the pivot line as the same weight would have if it was on the centre line.

The effect will be somewhat reduced by the increased overall weight of the trike relative to the base bike, but will still be substantial.

You could say that although mass concentration, (the positioning of all heavy components as close as possible to the centre of gravity), is a good thing on all performance vehicles as it aids agility (the ability to rapidly turn into and out of a corner), a three-wheeler is something of an exception.

Mass concentration in the fore and aft direction is certainly good but, across the machine, two concentrations of as much mass as possible towards the outer edges is a better solution on anything other than very smooth surfaces.

I hope the above makes sense to readers but getting back to the question of what spring characteristics do we need.

The springs need to provide a relatively high load at each of the front wheels similar to that of the base bike's forks when in the static position.

They must each also provide a much lower rate (lbs load increase per inch of compression) than the two fork springs combined.

This is where the moment of inertia about the pivot line, mentioned above is important. A single wheel bump produces an upward force on the wheel which must either compress the spring or momentarily lift that corner of the trike, causing a very rough ride. The moment of inertia will be much lower on a trike than on the base bike, as shown in Sketch 44, so the springs need to be softer to ensure that they compress and absorb most of the upward wheel movement.

To put some numbers to it, based on my experience, somewhere between 50 and 75% of the spring rate of the bike forks at each front wheel is optimum between too hard a ride and frequent bottoming of the suspension.

Spring Design

The first thing to say about finding the ideal front springs is that the first attempt is unlikely to be the last so adjustability in both spring rate at the wheel and the ride height are best designed in from the start.

Some ride height adjustability is often built into the suspension unit but some extra adjustment by moving the attachment of one end of the suspension unit can be very useful.

Spring rate at the wheel can usually be adjusted by moving one of the suspension unit ends, perhaps by a series location holes which change the ratio of wheel movement to spring compression.

I have used rear mono-shock suspension units from a bike of similar weight and similar wheel weight and ratio of wheel movement to suspension unit compression on my machines with different springs fitted.

It would have been good to have a shock absorber with adjustable compression and rebound damping but the standard non-adjustable rear unit I have used seems quite well suited to the trike front end.

The main purpose of the shock absorber is to resist rapid spring compression and rebound to avoid over-swings which would result in excessive bottoming on compression or start a bouncing motion on rebound. The resistance will basically increase with the square of the speed of compression or rebound, usually modified by internal restrictors, so it will have little effect when entering a corner as the rate of spring compression is relatively slow.

Back to the springs — The base bike rear suspension spring in my suspension gave a rate at each front wheel of 120 lb/inch which is far too high and resulted in a very harsh ride over small undulations although it absorbed "pothole" shocks well. After a lot of experimentation, I settled on variable rate springs with an initial rate of 60 lb/inch rising to approx. 120 lb/inch at the bump stop. This seems to work quite well on a machine with 180 lb sprung weight on each front wheel although there is always room for improvement.

The Rear Wheel

At first sight, the rear wheel suspension requirements seem little changed from the base two-wheeler with similar sprung weight on the wheel including the rider, but the same moment of inertia effect as described at the front also applies to some extent at the rear with the overall centre of gravity moved from mid-length to approximately 1/3 of the wheelbase from the front.

A reduced spring rate is therefore likely to be needed to achieve a similar ride quality.

Special Springs

It is likely that special springs will be needed as neither front nor rear springs from a two-wheel bike are likely to be suitable, so not getting it right first time can get expensive.

Spring design calculations are quite complicated as, although a coil spring is basically a torsion bar wrapped into a coil, the curvature and slope of the coils relative to the central line of force are complications affecting the final rate.

On-line spring calculators are very useful and there is a very good one on the web site of the US Company (acxesspring.com).

As a quick approximation if you have an existing spring, the spring rate reduces:

- *In proportion to an increase in the number of working coils*
- *In proportion to the square of an increase in coil diameter*
- *In proportion to the 4th power of a decrease in wire diameter.*

Not accurate but gives a quick idea of the sort of change which may be required.

Long compression springs will tend to buckle sideways and need support from an internal or external tube to keep them straight during compression. It is better to go for a larger outside diameter to reduce the rate of a spring than to increase the number of turns. If the ratio of free length to outside diameter is less than 4, the spring should be stable in compression without support.

For one-off spring manufacture, Active Springs at Studley, Warwickshire are a good bet but the spring details need to be very clearly stated. *(Spring manufacturers tend to count all turns including the non-moving end turns so the required number of working turns needs to be clearly stated.)*

Photo 36 shows my "final" spring in position on the Mk 2.

Photo 36 – The final front springs (after much experimentation).

3.3 – ANTI-ROLL BARS

Anti-roll bars are primarily intended to limit roll during cornering, and are particularly necessary when the roll centre is low.

They also have an instantaneous effect on behaviour on a single wheel bump, in that the anti-roll bar will absorb some of the initial single wheel upwards impact load, by transferring it to the other front wheel spring. This part of the impact load will then be shared between the two sides, tending to lift the whole front end rather than just the bump side corner.

The relevant "Moment of Inertia" when the whole front is lifted pivoting about the back wheel is much higher than that relevant to a single wheel bump so, all other things being equal, more of the impact load will be isolated from the rider.

This is another area where the best compromise between cornering and rough surface suspension operation needs to be found.

3.4 – BUMP STOPS

Good failure analysis practice requires independent rubber bump stops positioned so that they will support the machine with sufficient ground clearance if the spring mounting become detached of any reason.

I have used bump stops from the BL Mini as shown in Photo 37. A check with the spring unit removed showed that they compress about ¾ inch with a 3 to 1 ratio between wheel movement and the movement at the bump stop position and 180 lb of sprung weight on the wheel.

Photo 37 – The bump stop in position.

A Final Thought

The front suspension is one of the two key features which can make the difference between a machine which feels natural and is a pleasure to ride and one which does the job but needs constant vigilance to ensure control, particularly on rough surfaces. If you produce a satisfactory suspension design, you are halfway there. The other key factor is steering, which we will come to shortly.

Much time and effort will inevitably be saved if you build in from the start:

- The ability for some adjustment to the spring rate felt at the wheel without replacing the springs.
- Sufficient clearance to accommodate larger diameter springs if necessary.
- Additional ride height adjustment.

4 – Steering

The second key feature with two wheels at the front is steering, which can be separated into 6 separate aspects, in effect minimising heavy steering and 5 key angles.

4.1 – HEAVY STEERING

Heavy steering, particularly at low speed, can result from having only about 30° handlebar movement per side from lock to lock compared with two full rotations of a steering wheel on a car. It is therefore essential to avoid any additional unnecessary friction in the steering, in particular if the front end is connected to the bike at the steering head by the method shown in Sketch 41. A ball or roller bearing must be included at the load carrying connection to the steering head and all load carrying ball joints at the steering hubs must be as friction free as possible.

Based on good failure analysis principles, I have used car parts for all the load carrying moving joints between the handlebars and the wheel hub carriers as these are designed for much greater loads, are well proved in use, and are very well sealed and as friction free as possible.

The other factor which can affect steering weight is caster angle, the forward inclination of the wheel steering pivots (the equivalent of trail on a two-wheeler). This is necessary for stability but larger than necessary caster angles introduce additional resistance to the steering at very low speed because turning the steering makes the two tyre contact points move sideways by different amounts. This effect disappears at speed so is not a major problem and should not influence selection of the best caster angle for straight line stability.

The possibility of power steering, as is used by one of the few manufacturers of reversed trikes, is sometimes raised but I have avoided this approach because a satisfactory manual system is possible without the extra complication (and possible failure analysis concerns). "Keep it simple" is always a good motto when you are looking for safety and reliability.

The system I have used is to connect a lever attached to the existing fork tubes directly to steering arms on the wheel swivel hubs by simple track rods but, as with everything else in the steering system, getting all the angles right is the key. Sketch 45 shows the principle.

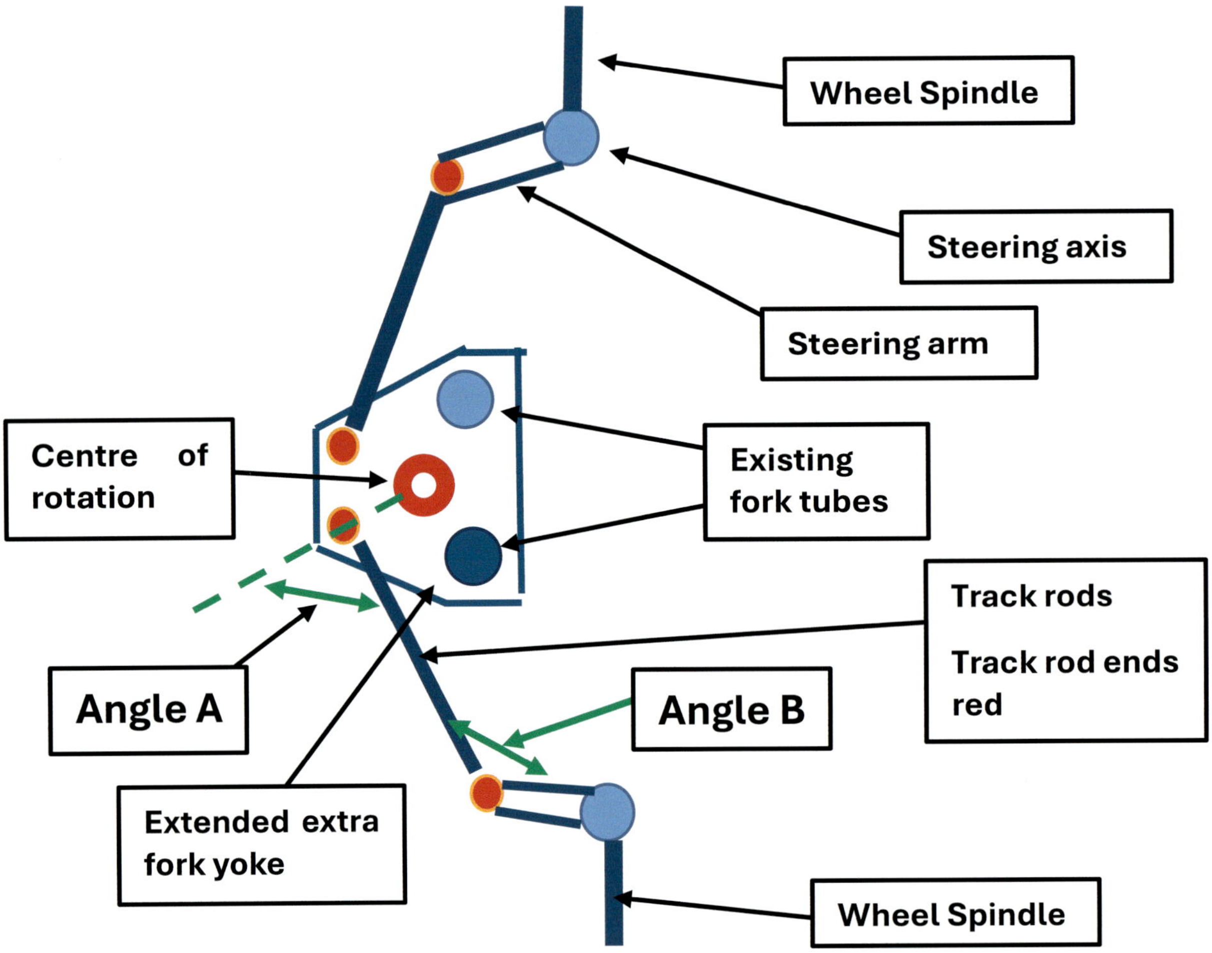

Sketch 45 – Steering arrangement (as used on my machines).

Sketch 45 is self-explanatory and is the simplest method of connecting the steering. The fork tubes do not need a bottom support bearing as they experience much higher loads under front wheel braking on the two-wheeler.

The extended extra yoke is a third fork yoke of the same type as the top yokes and provides an accurate centre of rotation. It is bolted to an aluminium plate extension to attach the track rod ends. These are old BL Mini parts which are well proven, better sealed than the typical ball joint, and relatively low cost compared with typical universal ball joints. The only disadvantage with them is that they are all right-hand thread ($^1/_2$ inch UNF) meaning that some additional means of adjustment is required for accurate final toe in setting. *(I included a right- and left-hand threaded sleeve fitted in one track rod.)*

Most of the system can be just arranged to fit the dimensions of the machine but there are two critical angles shown as angles A and B on the sketch.

> **Angle A** is the first key feature and must be 90° on both sides with the handlebars in the central position. This will ensure that the left and right-hand track rods move the wheel steering angles in the same manner whether the handlebars are turned left or right.
>
> **Angle B** The wheels need to turn by different amounts on the inside and outside wheels on a corner as the inside wheel runs round a smaller radius than the outer wheel and therefore needs to be turned through a greater angle. This is normally achieved by the setting of angle B (Sketch 45) which achieves the required effect if it is greater than 90°.

The best setting for angle B is complicated by the fact that, when loaded sideways on a corner, the tyres run at slip angles creeping sideways as they roll round the corner. Slip angles vary with cornering speed, road adhesion, tyre type and pressure, weight on wheel, power transmitted or braking. Sketch 46 illustrates the effect.

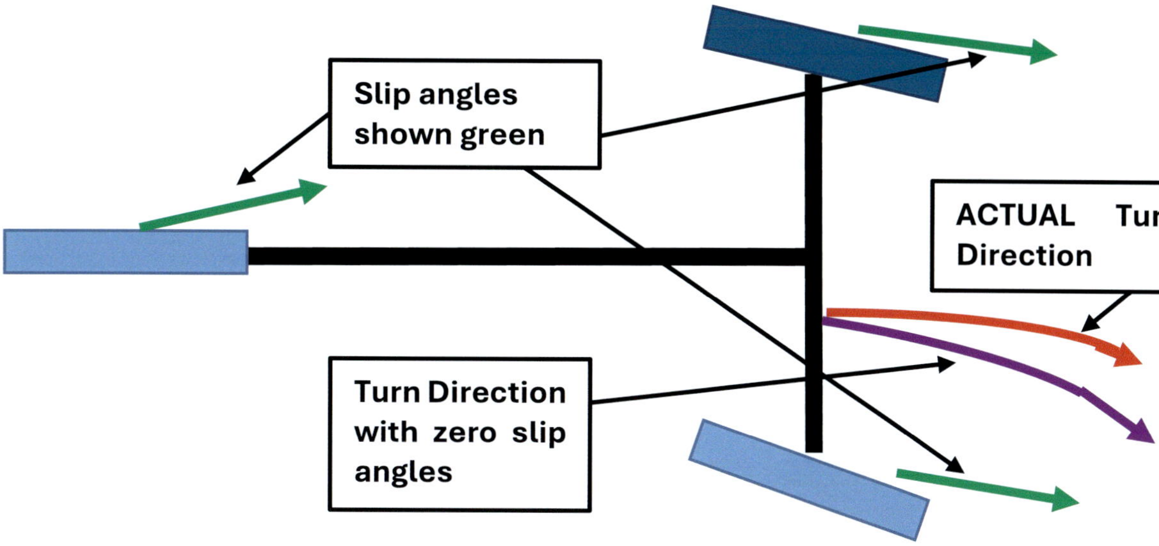

Sketch 46 – Showing how slip angles affect cornering.

Sketch 46 shows similar slip angles at both ends and a mild understeer condition resulting in the actual turn radius being larger than it would be with zero slip angles. This is a desirable set up with neither power nor braking applied. If power or braking was then applied at the back wheel, the rear slip angle would increase, moving the balance towards oversteer and tightening the turn radius. Front wheel braking on a corner would have the opposite effect.

Back to angle B. The factors mentioned above will vary over a large range during a single journey, so obviously angle B has to be a compromise.

With zero slip angles, the centre of the turn radius would be on a line through the centre of the back wheel, as shown in Sketch 47.1. For both front wheels to roll with zero slip angle, each front wheel would need to turn so that the purple dotted lines through both front wheel axles meet at a point on the dotted centre line through the fixed back wheel, marked as point X.

If we introduce slip angles, Sketch 47.2, the purple dotted lines move to a position at 90° to the actual direction of motion of each wheel along the line of each slip angle. For geometrically correct steering, the 3 purple dotted lines would need to meet at a point which would be further forward than the zero slip angle case.

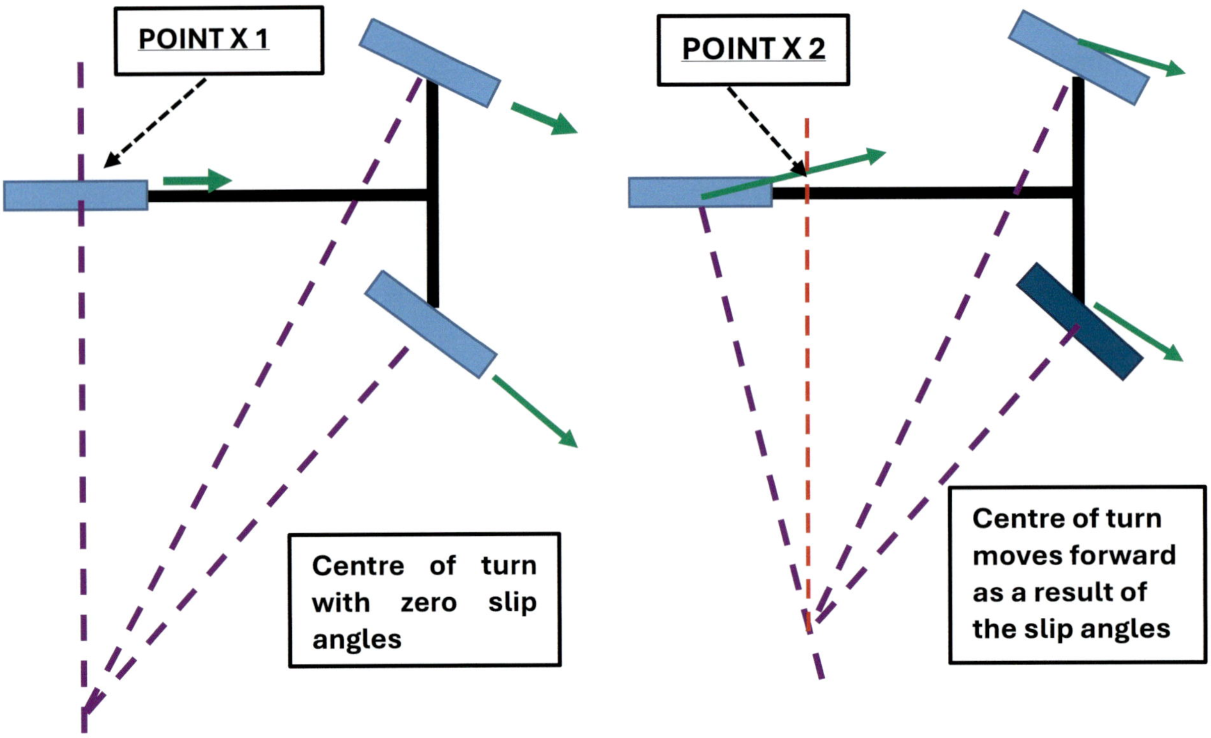

Sketch 47.1 (left) – Centre of turn with zero slip angles.
Sketch 47.2 (right) – Centre of turn taking into account slip angles.

The points marked X 1 or X 2 on the centre line are in line with the centre of the turn and the position of X 2 will move forward with an increase in slip angles.

This is where the compromise comes in as we need to select an average position of X 2 in order to find the best setting for angle B between the track rods and the steering arms at the swivelling wheel hubs.

For road use, position for X 2 at 1/3 of the wheelbase from the back is a good compromise and I have used this on both my machines with no obvious problems.

So finally, we get back to angle B. Sketch 48 shows the steering in the straight ahead position. The left side shows track rods assumed to be at 90° to the centre line as is necessary to find angle B. The right side shows the effect of angling the track rods because of the relative positions of the existing forks and the wheels.

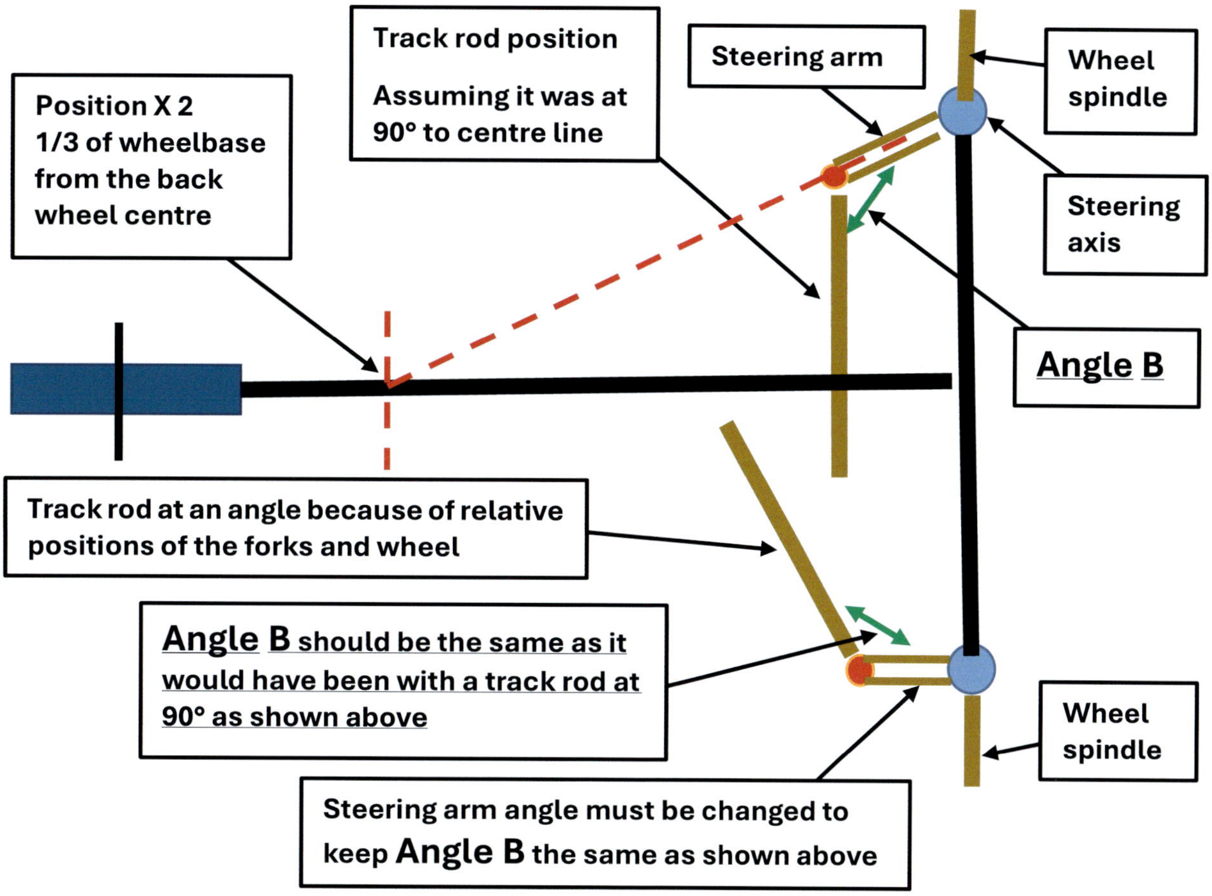

Sketch 48 – Showing steering arm angle B and the effect of an angled track rod.

A bit of a long-winded explanation but getting angle B right is a fundamental of good steering. If it is not close to that shown above, the two front wheels would be attempting to turn different corners with the final effect somewhere in between and the tyres scrubbing sideways with a very bad effect on both control and feel.

As a guide, on my Mk 2 machine ***Angle B*** *works out at 122° and I expect that about 120°would be satisfactory on most conversions from a standard 2-wheeler.*

The 3rd Angle which needs to be fixed at an early stage is the inclination of the steering pivot towards the wheel. The line through the steering pivot should hit the road as close as possible to the wheel centre line when viewed from the front. This ensures that forward or rearward forces on the wheel, for example from a single wheel bump, do not result in a large torque load at the steering axis which would be felt at the handlebars. Sketch 49 shows the effect.

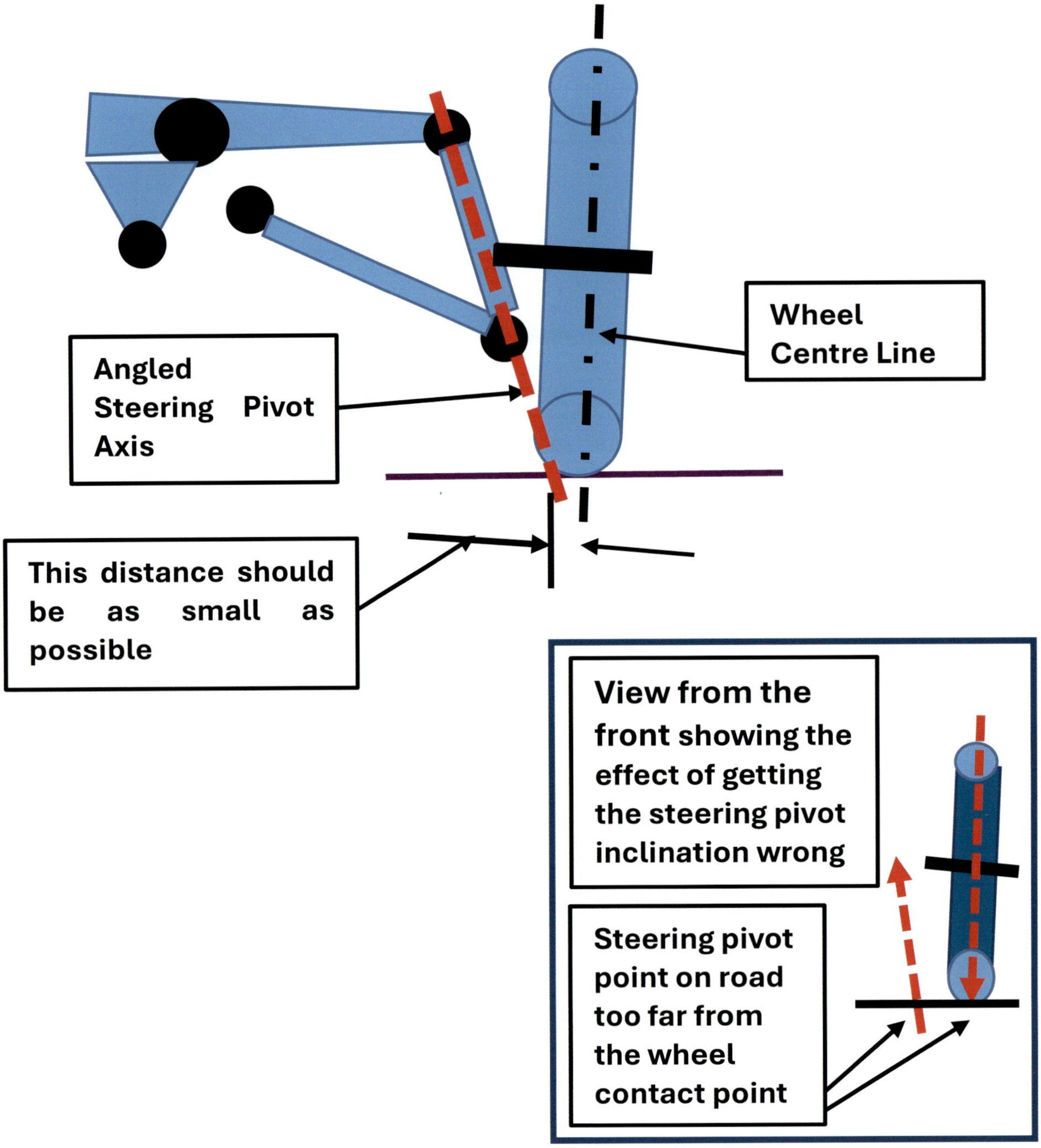

Sketch 49 – Steering pivot inclination.

The insert shows the effect of getting the steering pivot inclination wrong as shown by red arrows. Any horizontal impact force on the wheel at the tyre ground contact point will be a force momentarily turning the steering which will need to be resisted at the handlebars,

It is easier to achieve a close alignment as shown in Sketch 49 with a dished wheel from a car or two rear wheels from a bike with single sided swinging arm but is possible with a motorcycle front wheel by increasing the angle of the steering pivot.

I used two Honda VFR750 back wheels on my Mk 1 machine, which fit in well with the car type hub arrangement but on the second, I used two of the base ZXR400 type front wheels with one brake disc on each. This gives a much lower un-sprung weight but necessitates more outward slope of the steering pivot axis which hits the ground ½ inch inboard of the wheel centre line and seems satisfactory.

4.2 – STEERING DAMPERS

This seems like a good place to mention steering dampers. If all the various angles mentioned in this section are close to those suggested, there should be no need for a steering damper on relatively good road surfaces but on very uneven surfaces, especially if unexpected, I think fitting a damper is essential. Without one, off centre impacts on one front wheel which tend to momentarily turn the steering can only be resisted by the rider with the handlebars. Not a good experience, especially if there are a series of irregularly spaced ridges in the path of both front wheels.

I have only experienced something approaching a tank slapper once, on the Mk 2 machine without a steering damper fitted. It was caused by an unexpected series of potholes alternating between the two front wheels. The momentary forces generated at the handlebars were too high to be totally resisted by hand, even if the event had been anticipated, because of the low, almost one to one, ratio of wheel movement to handlebar movement. (With steering wheel steering there is usually at least one whole turn of the steering wheel each way.) The bike continued in a more or less straight line and settled down very rapidly, but it was not an experience to be recommended.

A hydraulic steering damper operates on the same principles as a suspension spring shock absorber and allows normal, relatively slow, steering movements with little resistance. If a momentary road impact on a front wheel results in a high turning torque on the steering, the damper will resist any rapid movement for the very short time until the impact is passed.

The best type of steering damper for a particular machine may need some experimentation but in general, it needs to be adjustable or considerably stiffer than the ones often fitted to two-wheel bikes.

Another point worth considering is that while most two-wheeler dampers are fitted in the steering head area, as it is the only place where steered and fixed parts are a suitable distance apart, this restriction does not apply on a two at the front three-wheeler. The steering damper is best placed as close as possible to the wheels, the source of the loads, while keeping it protected from road spray and debris. With the type of steering mechanism shown in Sketch 45, the best place is between the extra bottom yoke and the front sub-frame, wherever is convenient.

Not forgetting failure analysis, anything which could lock the steering is obviously a potential **Hazard.** Standard production dampers in general have a lot of service experience so can be assumed to be safe but the mounting position needs careful thought.

There should be plenty of clearance around the moving damper rod to avoid any possibility of fouling on static parts. If one end became detached of some reason, the loose end should fall free with no possibility of interfering with the steering.

I have used an adjustable bike damper which seems about right, to date, for my Mk 2 machine. It is mounted under the extended bottom fork yoke and at a suitable strong point on the main connecting tubes to the front sub-frame where it is below and well clear of all the other moving steering parts as shown below. It is positioned approximately parallel to the track rod and so moves directly with the steering and is behind the sub-frame giving it some protection from road spray and debris.

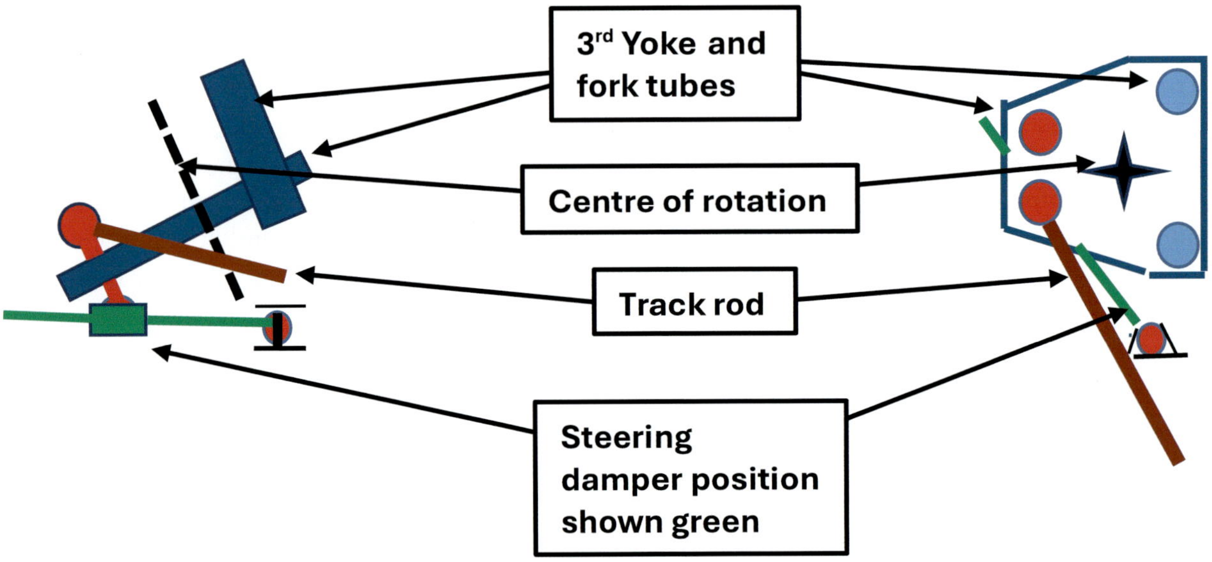

Sketch 50 – A good position to mount a steering damper.
The sliding attachment of the damper is mounted below the inner track rod ball joint and the fixed end to the static structure.

4.3 – SET UP ANGLES

The above three angles fix the geometry of the steering system and so must be settled early as they involve the design and making of parts but the other two, caster angle (the equivalent of trail in motorcycle terms) and toe in can be made easily adjustable on the finished machine to achieve the best overall stability and feel. Sketch 51 shows these angles.

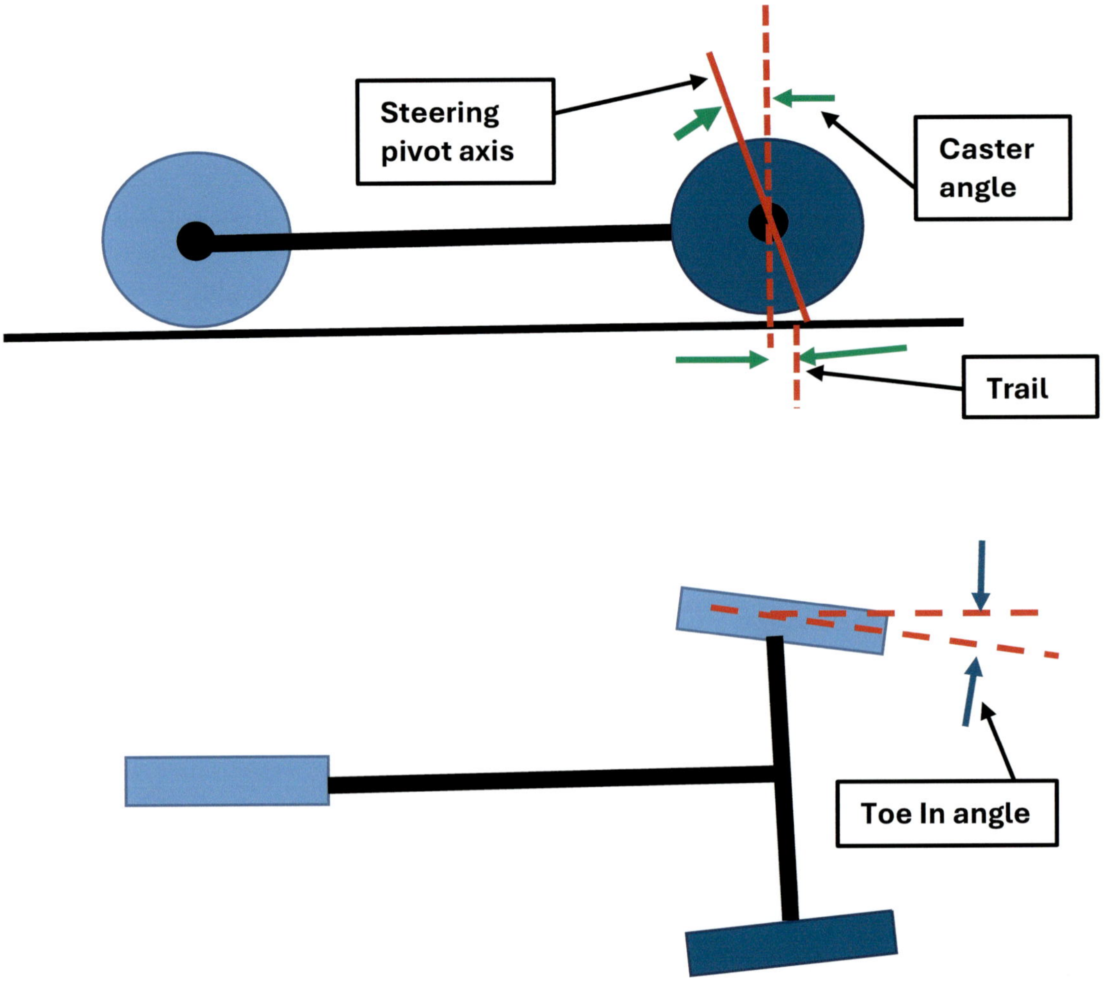

Sketch 51 – Caster angle and trail dimension (top). Toe in angle (bottom).

Caster Angle

Caster angle/Trail is the most important and should be set first. It sets the point at which the steering axis meets the road surface. This point must be in front of the centre of the tyre contact patch which is vertically below the wheel centre. This dimension is equivalent to the trail in motorcycle terms and provides the self-centring effect which makes the steering always tend to move to the straight ahead position when moving forward on a smooth surface.

The best caster angle and trail for a particular machine are subject to some experiment. Too much will provide straight line stability but may introduce very heavy steering. Too little will result in the need for constant attention to maintain a straight course.

On my Mk 2 machine, after much testing, I have settled on a 10° caster angle giving approximately 2 inches of trail with a 24-inch rolling diameter. This gives good stability on most surfaces without excessively heavy steering, but all machines are different, and adjustability is the key to achieving the best settings.

On a handlebar steered machine on extremely potholed surfaces, no amount of trail will ensure straight-line stable running without assistance from a steering damper, ref section 4.2, but a satisfactory set up should ensure rapid return to self- centring.

Toe In

It is normal to set both front wheels so that they run at a very small slip angle turned inward relative to the direction of straight line motion. Each wheel therefore produces a small inward force on the machine. On a smooth surface, these forces are equal and have no effect on the direction of travel. A bump impact on one wheel will then momentarily increase the inward force generated by the fixed slip angle pushing the front end inwards away from the bump side and countering any tendency for the bump impact to cause a turn towards the bump side.

A small amount of toe in is therefore beneficial but must be set to the minimum necessary as the continuous slip angles are not good for tyre wear.

I am currently using a setting equivalent to ¼° on each wheel which is approximately equivalent to 0.1 inches across each 24-inch tyre diameter. That is a measured toe in between wheels of 0.2 inches (using the old straight edge and measuring tape method.

One point worth noting is that in order to ensure that the machine runs straight with the back wheel and handlebars central, the first thing to set is the alignment so that, with both front wheels parallel and the handlebars in the centre position, the back wheel is central relative to the front wheels as shown in Sketch 52.

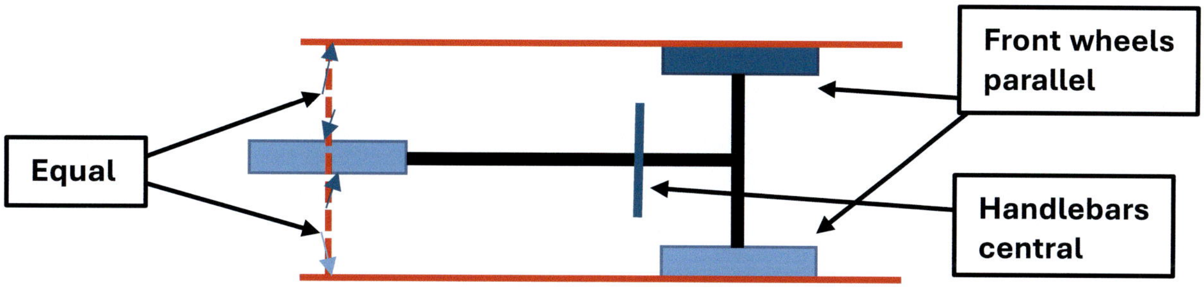

Sketch 52 – Alignment check before setting toe in.

With the alignment correctly set, toe in can be adjusted by equal adjustments to the length of the two track rods as shown in Sketch 53.

(Minor errors in alignment are not really noticeable when riding but following riders seem to enjoy telling you that the back wheel is not running in the middle.)

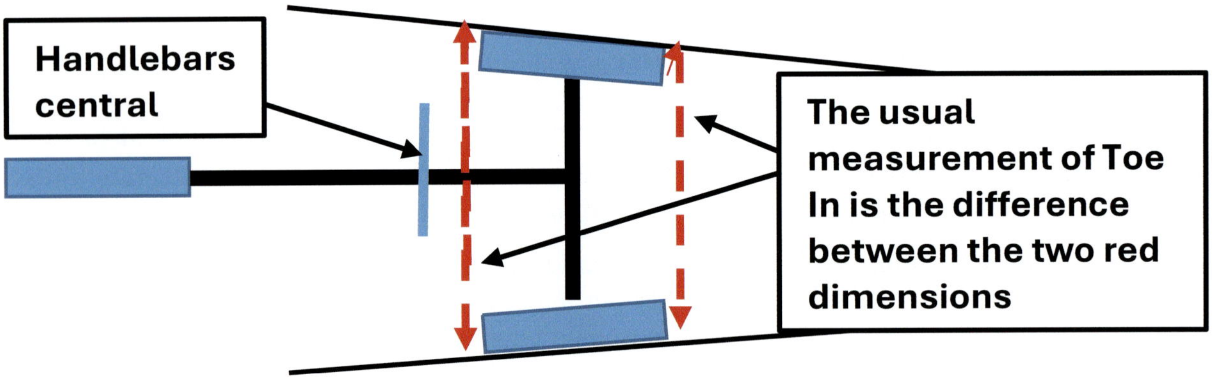

Sketch 53 – Setting toe in from the aligned condition.

More modern 4-wheeler methods use lasers for straight line references but may have to be adapted to cope with three wheels. The old world method, measuring dimensions between long straight edges aligned against the tyres, always works and is foolproof and sufficiently accurate.

- *With the handlebars approximately central, set the 2 front wheels approximately in the straight ahead position and exactly parallel to each other using the track rod adjusters*
- *Turn the handlebar slightly until one front wheel is aligned parallel to the back wheel and measure between the two parallel straight edges held in contact with the tyres.*
- *Without moving the handlebar, repeat on the other side. The measured dimension between the two parallel straight edges should be the same on both sides but a small difference (indicating that the rear wheel is not running exactly central between the two fronts) is inevitable on a machine constructed without a rig and will have no noticeable effect in use.*
- *Reset the handlebars central by equal adjustments in opposite directions to the two track rod adjusters without turning the front wheels.*
- *Then set the required toe in by equal small adjustments on both track rods.*

(If your straight edges are not guaranteed to be perfectly straight, turning them over and repeating the measurement, then taking the average of the two, will give an accurate dimension)

Adjustability

Toe in, caster angle (trail) and ride height will inevitably need some adjustment during road testing, so it is better to build in easily adjustable features from new.

Photos 38 and 39 show caster angle and ride height adjusters on my Mk 2 machine with toe in adjusted in the normal way by means of threaded track rod ends.

Photos 38 and 39 – Caster angle toe in and ride height adjusters on the Mk 2 machine.

5 – Wheels

There are a large variety of wheels which could be used on a reverse trike but there are a few essential features, particularly at the front, needed to achieve satisfactory steering and suspension.

5.1 – REAR WHEELS

If we start at the back, there would seem to be no advantage in using anything other than the existing motorcycle wheel. From the point of view of strength with the different cornering side loads on a trike, there should be no problem as maximum side loads are limited by tyre adhesion and will be similar to the side loads on the wheel which result from severe bump loads at high lean angles on a two-wheeler. Occasional impacts with kerbs, etc. will also be similar to the two-wheel situation.

There is also some advantage relative to UK registration in minimising change to the original bike.

5.2 – REAR TYRES

Rear tyre wear can be a problem with enthusiastic cornering as large slip angles are easy to generate and very controllable. I have found that the best rear tyres for a motorcycle based reversed trike are motorcycle rears designed for heavy cruisers. They have good wear characteristics when upright without too large a contact patch.

(The wet roundabout is one place where a reverse trike excels. It would take a very determined rider to match one on the equivalent two-wheeled bike.)

Many people have suggested car tyres for the back wheel, and they would cope well with respect to wear, but I have avoided them as the larger contact patch with only around 200 lb weight on the wheel would tend to promote aquaplaning in the wet.

5.3 – FRONT WHEELS

There are a large variety of possible front wheel choices but, whatever wheels are chosen, there are three aspects which must be right if a satisfactory steering and suspension arrangement is to be achieved.

i) The wheels, together with all un-sprung steering and suspension parts, must be as light as possible to achieve the maximum possible ratio of sprung to un-sprung weight. This makes it much easier to find satisfactory front suspension settings.

ii) The combination of wheel and steering parts must allow close proximity between the steering axis and the centre of the tyre/road contact patch as shown in Sketch 49.

iii) The tyre profiles must provide a contact patch area (footprint) small enough to avoid aquaplaning in wet conditions and reasonably constant over the range of suspension movement.

The first requirement is easiest to achieve by using motorcycle wheels and brakes with the brake disc mounted on the wheel, but this means that the second requirement is less easy, but not impossible, to achieve. The existing motorcycle wheel bearings in the wheel can be used providing that the spindle diameter is large enough to allow single sided mounting on a fixed stub axle.

Smaller wheel diameter would obviously reduce the weight of the wheel but may look a bit odd. I have stuck with the same wheel diameter as the base bike because appearance matters on a project of this sort, and I think same wheel size at each end looks more balanced. Another decision to be made early in the project.

I used front wheels from a Kawasaki ZX 9R on my Mk 2 machine which are the same spoke pattern as the original ZXR400 wheels but have a larger 25mm spindle diameter. An alternative may have been to find some suitable bearings with the same outside diameter as the bearing locations in the ZXR400 wheels but with a larger inside diameter. These parts are all safety ***Critical*** *so it would be safest to stick to well proved car or trailer bearings.*

Photo 40 shows the wheel mounted on the Metro swivel hub with the brake disc replaced with an aluminium pointer to allow more accurate clearance checks, minimum running clearance between rotating and static parts being necessary in order to satisfy requirement (ii) (a minimum clearance of 0.125 inches has proved to be sufficient.)

Photo 40 – Showing a front wheel mounted on the metro swivel hub carrier. The second 25mm stub axle is shown below.

If the base bike has a front wheel driven speedometer, this introduces another complication as the space it occupies in its original position would move that wheel outwards, making the second requirement more difficult, if not impossible, to achieve. I overcame this by moving the speedometer drive to the inner end of the sub axle and by running on the car wheel bearings in the Metro hub carriers with the bike wheel locked to the stub axle by means of a locking plate driven from the unused outer brake disc brake disc mounting threaded holes. Photo 41 shows the arrangement.

Photo 41 – Speedometer drive arrangements on the Mk 2 machine.

The 5 studs at the former brake disc positions drive an aluminium disc which is clamped by the main stub-axle nut and drives through a short hexagon section cut on the end of the stub-axle. The base bike speedometer drive is then mounted on the inner end of the stub-axle.

Another small complication is finding a satisfactory route for the speedometer drive cable, together with the brake pipe and electrical connections, as they need to accommodate both steering and suspension movement without excessive tension, bending or fouling on static and moving parts. The best solution seems to be to run the cables forwards from the swivel hub carrier curving round 180° to attach to the static parts in a rearward direction. 7-core caravan connection wire is ideal for the electrical connection if one is required in an open wheel design.

The alternative to motorcycle type front wheels is to use bolted on dished wheels, either from a car or two rear wheels from a motorcycle with a single sided swinging arm. This makes the second requirement easy to achieve but is inevitably much heavier than the motorcycle wheel alternative because it necessitates using a car type swivel hub carrier with its wheel bearings and its heavy brake discs and all associated heavy parts.

On my first machine I used two Honda VFR750 back wheels with BL Metro hubs and steering parts. The Metro hub flanges were re-drilled to suit the wheel bolting positions and the Metro brake disc thickness was reduced in weight somewhat to allow use of Suzuki Bandit lightweight callipers in place of the very heavy car callipers. The arrangement gave no problems, but the total un-sprung weight was approximately 50% greater than the second machine, making satisfactory front suspension even more difficult to achieve. Photo 42 shows the arrangement.

Photo 42 – The front brake arrangement on the Mk 1 machine using all car parts except for the reduced disc thickness and Suzuki calliper.

The motorcycle suspension units attach to the car damper attachment points part way along the top arm of the double wishbone suspension.

5.4 – FRONT TYRES

Tyre choice is another decision to be made, but should achieve the best contact patch size for the relatively small weight on the wheel. I believe motorcycle tyres are best. Their relatively large cross section relative to the contact patch size allows more compliance over small road bumps which is beneficial to suspension. They also allow changes in wheel lean angles with suspension movement at the front without affecting contact patch area which can be beneficial to high roll centre suspension designs. With flat tread profile car tyres, the suspension design needs to minimise wheel lean to avoid running on the edge of the tread which limits the options for high roll centre designs.

Tyre wear is generally good at the front, but one characteristic is that wear patterns tend to be a bit one sided, similar to those which result from too much toe in, even with virtually zero measured toe in. I believe that this is because weight transfer to the outside wheel during cornering is greater than on most four-wheelers. This means that most of the front tyre wear occurs when they are on the outside of a corner with little or no load or wear on the front tyre of the inside wheel.

6 – Brakes

If we again start with the back brake, there is no good reason for changing the existing brake disc, but the situation is made more complicated by UK rules which require a hand (parking) brake. Separate additional mechanically operated handbrake callipers are available, but I have found them to be not very satisfactory and it is better to replace the existing calliper with one from a small car with a built-in handbrake. There are many types available and one with approximately the same hydraulic piston diameter will operate from the existing pedal and master cylinder. A car type handbrake lever can then be used with a simple cable connection. The calliper can be usually mounted from existing calliper mounting bolt positions with a suitable adapter plate. The only real disadvantage of this arrangement is that car parts are usually a lot heavier than they need to be although lightweight versions are available for some cars at very high cost.

I settled for Citroen Saxo callipers on both the Mk 1 and Mk 2 machines with the pair conveniently covering both as the discs are on opposite sides of the wheel.

On the Mk 1 Honda Bros based machine, the calliper is at the top of the disc so the larger car calliper could be close to the same position but on the Kawasaki based Mk 2 machine it was at the bottom of the disc with insufficient ground clearance for the same type of car calliper. A suitable curved mounting plate cut from 10mm aluminium alloy plate and attached to the existing calliper mounting position moved the calliper round to the back of the disc and aligned it with the rotating disc. The original brake torque link was also attached to the mounting plate in its original position. Both arrangements have been problem free to date. Photo 43 shows the car calliper in position on the Mk 2 machine. It looks quite bulky on the photo but in practice it does not look out of place.

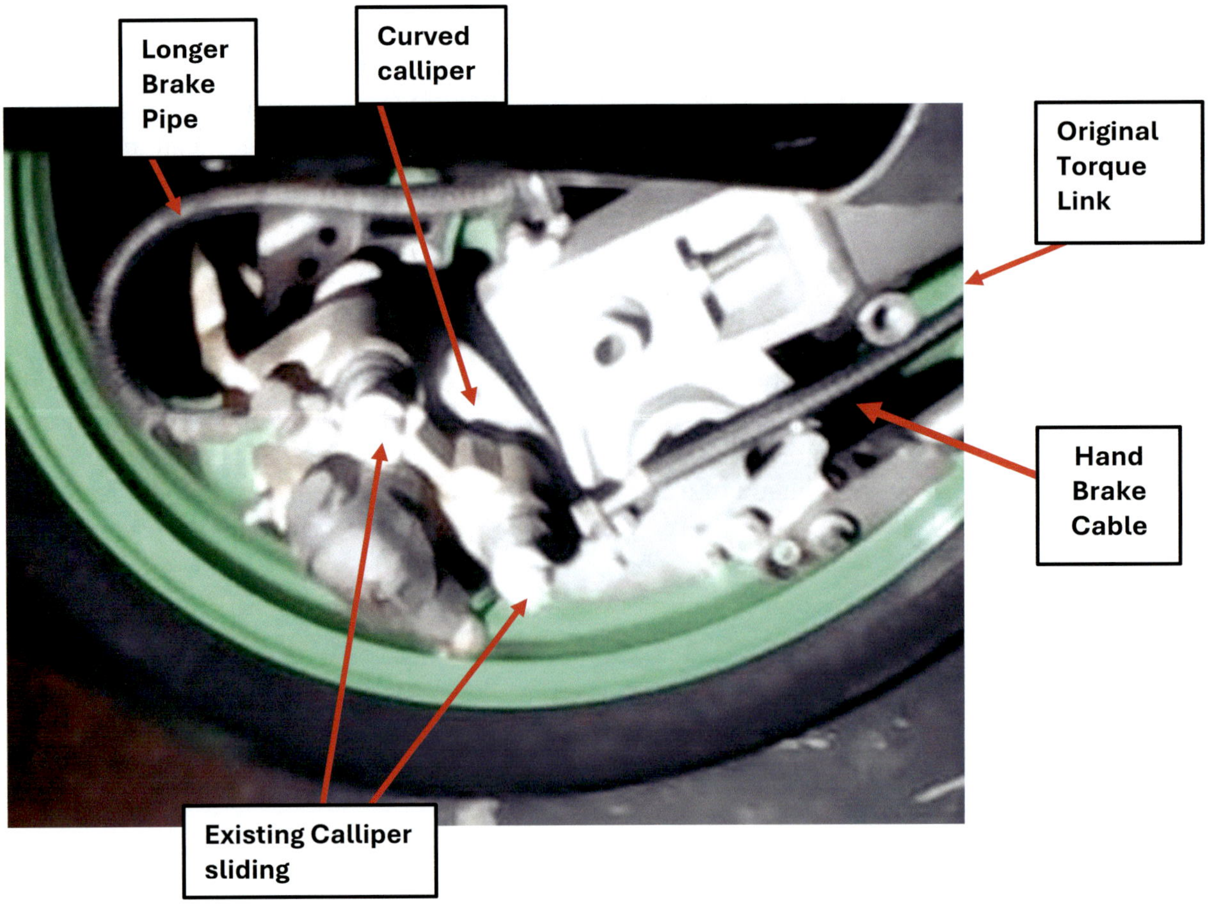

Photo 43 – Citroen Saxo rear brake calliper in position on the Mk 2 machine.

Photo 44 shows the MG handbrake lever which fits well into the ZXR 400 side panels and is in a good position for use. It is however only really there to satisfy the construction and use and MOT test requirements and is rarely, if ever, of much use in normal road-going conditions with motorcycle type controls.

It cannot be used for hill starts as both hands are in use for the clutch and throttle and the foot brake can be used to prevent roll back if necessary. It is also not useful as an emergency brake if the front brake fails, as the foot brake would be more effective on the back wheel. In fact, its only real use is as a parking brake where leaving the bike in bottom gear is normally all that is necessary.

The hand brake lever on the Mk 2 machine

The lowered and increased width foot rests on the Mk 2 machine

Photos 44 and 45 – The handbrake lever and footrest arrangements on the Mk 2 machine.

The existing brake pedal pivot moved down with the left footrest mounting so the lever only needed adjustment of its angle to suit the new footrest position, The gear lever shown above needed a longer connecting rod with the lowered footrest.

7 – Cooling System

7.1 – COOLING WATER

Assuming that the base motorcycle is water cooled, there are two alternative strategies.

The easiest system is to leave the radiator in its original position and make no changes to the cooling water system, but the front bodywork could easily block airflow to the radiator so ducted air through the bodywork is likely to be necessary. These air ducts tend to use up too much of the space in the front compartment, restricting its use as a storage space.

I believe that the best option is to move the radiator and cooling fan to the front of the bodywork where it has unrestricted airflow. Pipework to connect the cooling system to the radiator occupies very little space.

If the base bike system has an inaccessible pressure cap hidden under the fuel tank, the extra pipework provides an opportunity to include an accessible second small header tank with a sealed cap, as shown in Sketch 41 in Part 3, to allow easy water level checks which can be very useful during initial testing.

It is also a good idea to incline the radiator forwards at the top which allows convection cooling when stationary in traffic.

Photo 46 shows the radiator installation on the Mk 2 machine. The radiator was the one removed from the Honda Bros during its conversion to the Mk 1 and proved ideal for the Mk 2. It is considerably smaller than the ZXR 400 standard radiator but, with its better air flow and inclination, it has proved sufficient for road-going use. The cooling fan, also ex Honda Bros, has never switched on to date even in traffic in very hot weather.

Photo 47 shows the second header tank on the Mk 2 machine.

Photo 46 – The Mk 2 machine with front bodywork removed showing the inclined front mounted radiator.

Photo 47 shows the second header tank fitted at the same height as the original pressure cap, (inaccessible under fuel tank), and linked by pipes both below and above water level. (These tanks are sold as overflow tanks for kit cars but are very nice stainless steel tanks ideal for the job.)

The cap is sealed when closed leaving control of water pressure when running to the original under tank cap.

The Mk 1 took the convection cooling idea to extremes and mounted a small area three-core radiator horizontally at the front of the floor of the front compartment with cooling air entering below. It worked well but was perhaps too much of a good thing as it restricted other uses of that part of the front compartment.

Photo 47 – The second header tank on the Mk 2 machine. Useful for water level checks or top up.

7.2 – OIL COOLING

If the base bike has an oil cooler, the same choice of whether to move it to the front of the bodywork or duct air to it needs to be made. As the oil cooler will be much smaller than the water radiator, I think the best choice is to leave it close to the engine and duct or scoop some cooling air from underneath the bodywork. Making no or minimal change to the original oil system is the safest way to avoid any problems in this area.

On the base bikes of both the Mk 1 and the Mk 2 machines, the oil cooler was mounted low down below the radiator, so it was left in that position and just tilted forwards on the Mk 2 to promote some convection cooling when stationary. The Mk 1 has a narrow chord scoop aerofoil projecting just below the front floor level and the Mk 2 has a wide flat scoop projecting only ½ inch below the front floor. Both have provided enough cooling air to avoid any problems to date and do not seem to collect too much road surface water or road debris.

Sketch 54 shows the oil cooler air scoop on the Mk 1 machine.

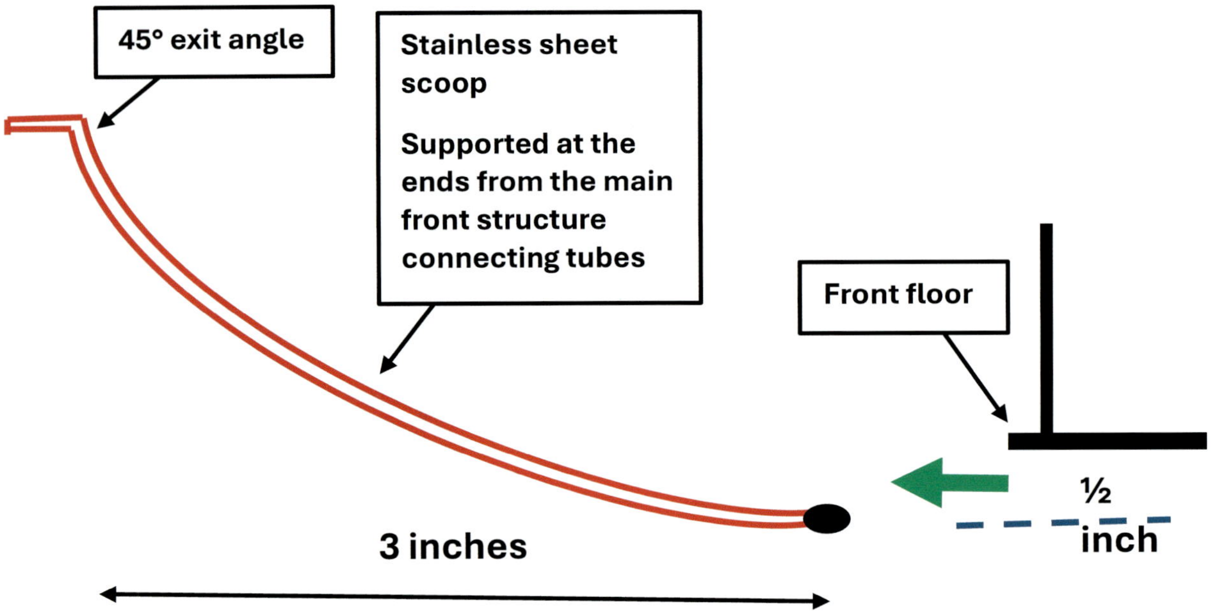

Sketch 54 – The simple cooling air scoop on the Mk 1 machine.

This simple sheet stainless steel scoop arrangement proved to be as effective in providing cooling air to the oil cooler and the area in front of the engine as the much more complicated and heavier system on the Mk 2. (There is always a temptation to make things more complicated than they need to be in search for some possible improvement. Simplicity is usually the best option.) It is often said, with some justification, that the search for perfection rather than settling for best achievable within time and cost constraints always leads to failure.

7.3 – EXHAUST COOLING

Exhaust pipes tend to run much hotter, particularly those in front of the engine, when shielded from the direct cooling air flow by the front bodywork. It is best to deflect some air from under the front floor into the area around the exhausts whether an oil cooler is fitted in that position or not. The warmer air then escapes from the top into the area in front of the rider, a small advantage on a cold day.

The air deflectors fitted for oil cooling on both machines also restored the air circulation in the exhaust pipe area in front of the engine and returned the exhaust pipe temperature to normal. A great improvement from a very small simple change.

Sketch 54 above shows the air deflector on the first machine.

8 – Electrical Changes

The changes needed to the electrical system are largely those necessary to comply with the various construction and use rules for light three-wheeled vehicles. I have tried to keep changes to the base bike electrical system to a minimum so as not to introduce any new problems.

The required changes are:

8.1 – HEADLIGHTS

Provided that the overall width of the machine does not exceed 1.3 metres it is permissible to keep a single headlamp. If it is wider, two headlamps are required. The positions of the two lights are defined in the IVA rules and it is best to check directly from the web site to avoid any problems later.

https://www.gov.uk/vehicle-approval/individual-vehicle-approval

In either case, the base bike headlight circuit requires no change other than extending the leads to the original headlamp to the new lights if required.

Both my machines are just under the 1.3 metres width so no change to the headlights was needed.

8.2 – POSITION LIGHTS (SIDE OR PILOT LIGHTS)

Additional side lights and rear lights are required at the outer edges of the wide front end and the existing rear light including the number plate light needs to be retained making a total of 5 position lights. The original pilot light in the headlight is not strictly required but may as well be left unchanged.

It is best to use LED bulbs in the new lights to keep the current demand similar to the base bike. The position of the new lights is defined in the IVA rules as above.

The other complication is that the rules require the position lights to be available with the ignition off. Most base bikes will need the ignition on to use the position lights so some change to the base bike wiring is necessary.

Most "relatively" modern bikes, including the Honda Bros in the Mk 1 machine, have a separate position light, (ignition on), positive feed wire to a handlebar multi-function switch independent of the headlight feed to the switch or that of the other switch functions.

In this case, it is only necessary to cut the (ignition on) position light feed wire to the switch somewhere along the handlebar, insulate the (ignition on) feed end and provide a new (ignition off) positive feed to the switch end of the cut wire. Then the existing handlebar switch will still operate the position lights with ignition off even though all other functions of the switch remained, ignition on, only.

The Kawasaki used in the Mk 2 Machine had a somewhat different circuit arrangement and, in this case, the easiest solution was to cut the wire from the handlebar switch to the position lights, insulate the handlebar switch end, and provide a separate position light switch on the steering head with an (ignition off) positive feed.

In both cases I used a separate small fuse box in the new front compartment to provide, ignition off, power supplies to both the position lights and the Hazard lights. Photo 50 shows the extra electrical panel on the Mk 2 machine.

The Sketches 55.1 and 55.2 illustrates the change.

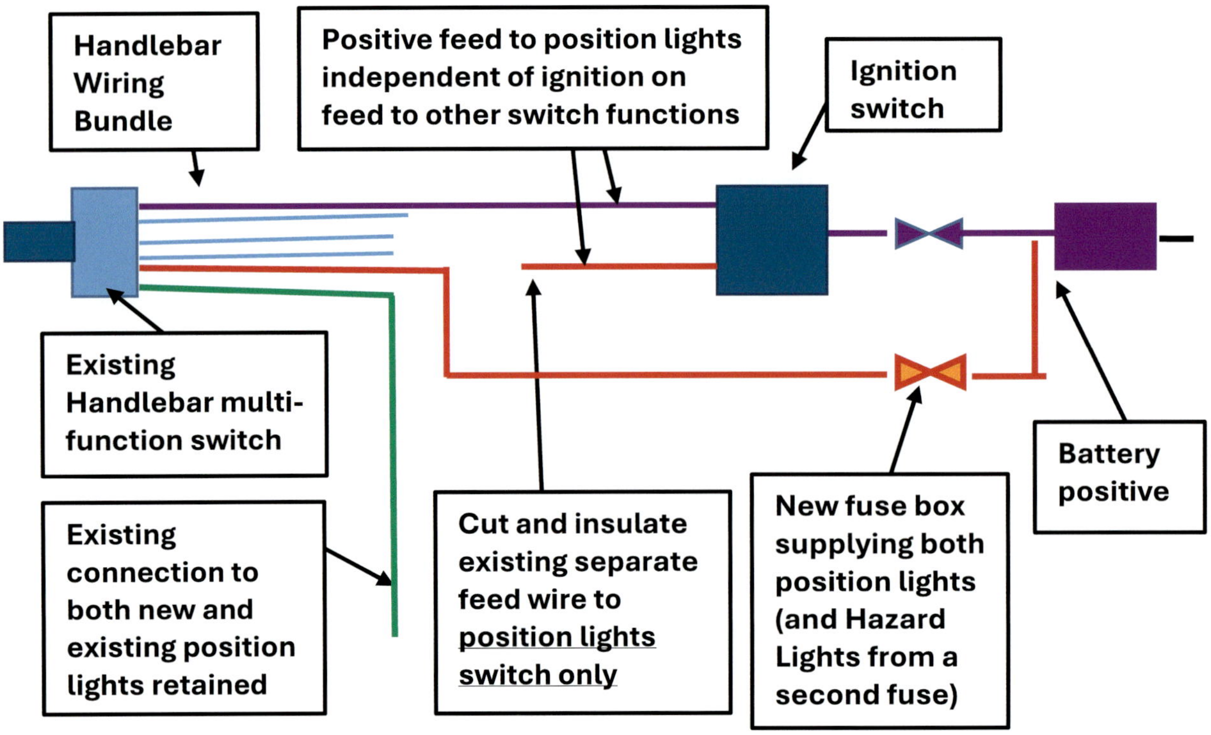

Sketch 55.1 – Suggested minimal wiring change if the position lights have an existing separate positive feed from the ignition switch.

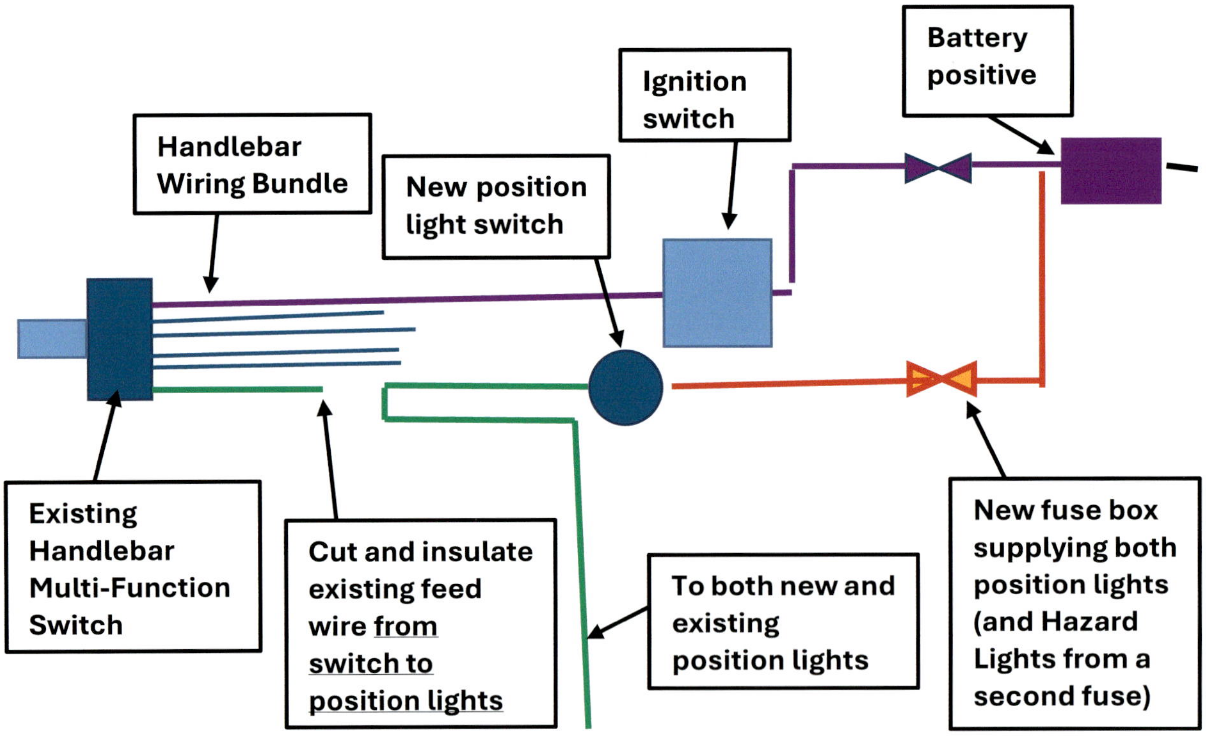

Sketch 55.2 – Suggested minimal wiring change if there is a single positive feed to the handlebar switches from the ignition switch which supplies both the position lights and other switch function.

Photo 48 in the brake light section shows the new lights including indicators and brake lights which are kit car equipment and are available from "Car Builder Solutions" who are very helpful and stock a wide variety of useful parts.

Photos 49 and 50 in the Hazard light section show the new switch position on the Mk 1 machine and the new extra fuse box in the front compartment of the Mk 2 for all the new electrical equipment.

8.3 – BRAKE LIGHTS

Extra brake lights are required with the position lights close to the outer edges of the new front section and it is easiest to combine them in a single unit with the position lights and indicators. Many varieties of kit car light units are available and are the easiest way to cover this requirement. IVA rules specify positions of the lights within the same limits as the position lights.

The easiest way to connect the extra brake lights is to run an extra wire to the rear of the base bike and connect into wiring in the final feed to the bike rear brake light which is usually easily accessible. Again, LED lights are best to keep the total current close to the original bike.

Additional rear reflectors are also required close to the outer edges of the new front end.

Photo 48 below shows the combined rear, brake and indicator light unit and the rear reflectors on the front mudguards of the Mk 2 machine. The front indicators are also on the mudguards, but the front position lights are on the front compartment lid. All are positioned within IVA rules.

Photo 49 in the Hazard light section shows arrangement of extra switches on the steering head.

Photo 48 – Showing light positions on the new front of the Mk 2 machine.

8.4 – INDICATORS AND HAZARD LIGHTS

Extra front and rear indicator lights are required close to the outer edges of the new front section which would be just a case of connecting them into the existing indicator circuit except that it is complicated by the need to fit Hazard warning lights which must be available with "ignition off" although the indicators are normally only connected with "ignition on." This requires a separate "ignition off" feed from the battery through another fuse in the new additional fuse box in the front section referred to for the position lights. It also needs some means of isolating all the base bike's "ignition on" equipment from the Hazard light feed when they are selected with ignition off. The easiest way to do this is to use a Hazard switch kit *(available as a kit from Car Builder Solutions)* which includes a special Hazard relay and switch with wiring instructions.

The UK requirements in the IVA rules specify the position of indicator and Hazard lights within the same limits as the position lights.

I never seem to do things the easy way and I used a system based on four diodes (which allow current to flow in one direction but not in the other), as shown in Sketch 56, with LED lights at the four front end positions and conventional bulbs in the base bike's existing rear lamp. It is best to use an electronic Hazard flasher relay for the Hazard system as they are less sensitive to total bulb current than the conventional type. The existing base bike indicator relay will then still operate the indicators separately when the ignition is on and the Hazard lights off.

The only change to the base bike wiring is to cut the two output wires from the indicator switch and extend the four cut ends as shown in Sketch 56.

This system works well but it is best to mount the diodes all together attached to a small piece of insulating board as they usually have single strand wire connections which would tend to break if included in a flexible wiring loom. (Miniature cable ties and a very small blob of silicone have been effective in fixing the diodes to the board.)

Sketch 57 shows the general idea.

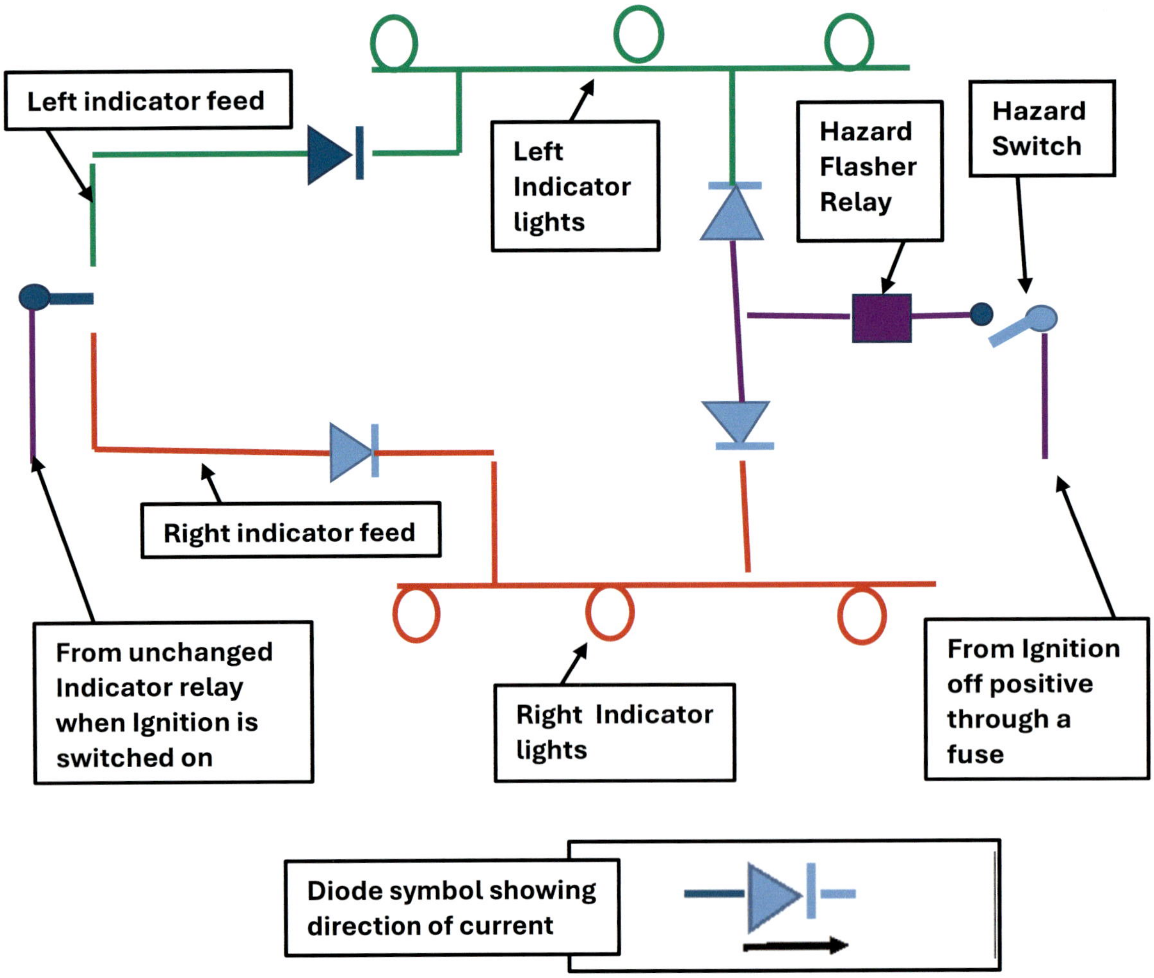

Sketch 56 – Showing circuit I used for the indicators/hazard lights.

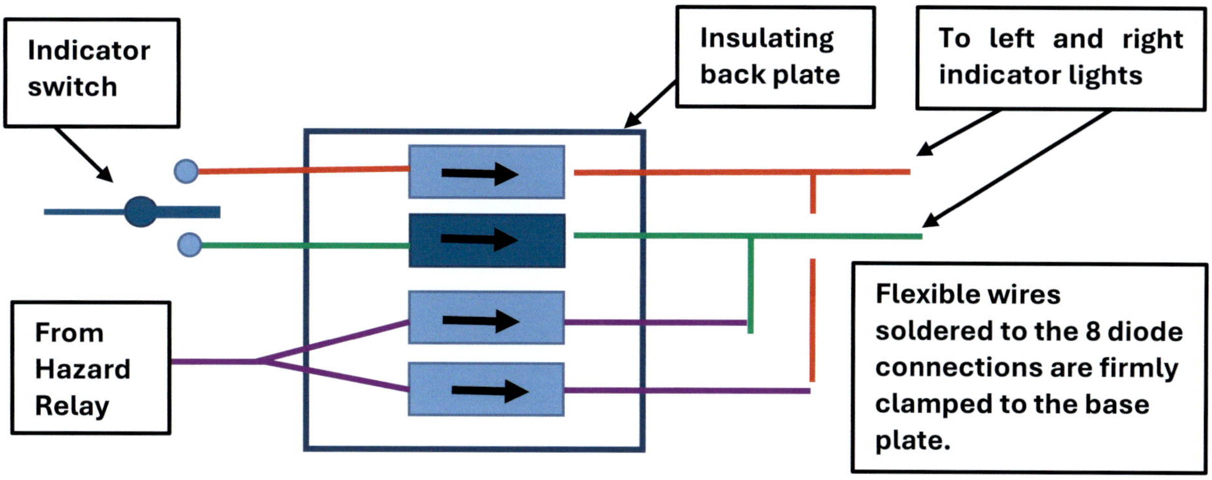

Sketch 57 – Showing the method I have used to mount the diodes for both indicators and hazard systems as shown in Sketch 56. If an indicator alert buzzer is included, a 3rd pair of diodes would be needed.

8.5 – INDICATOR ALERTS

Too much car driving tends to make you forget to cancel indicators after a corner, so some form of alert signal is useful to remind you.

I have fitted a buzzer which sounds when indicating but not when Hazard lights are operating. It has proved to be useful but needs to be quite loud to be heard through a helmet in traffic. It would be better if the sounder was on the helmet, either plugged into the bike or perhaps with a wireless connection.

The other alternative would be a bright flashing light on the steering head or in the front fairing, if one is fitted, but it would have to be shielded from all directions except towards the rider and would need a dimmer resistance coming in when the position lights were on.

A suggested circuit for an indicator alert, which is to some extent the opposite of the Hazard light circuit, is shown dotted in Sketch 58.

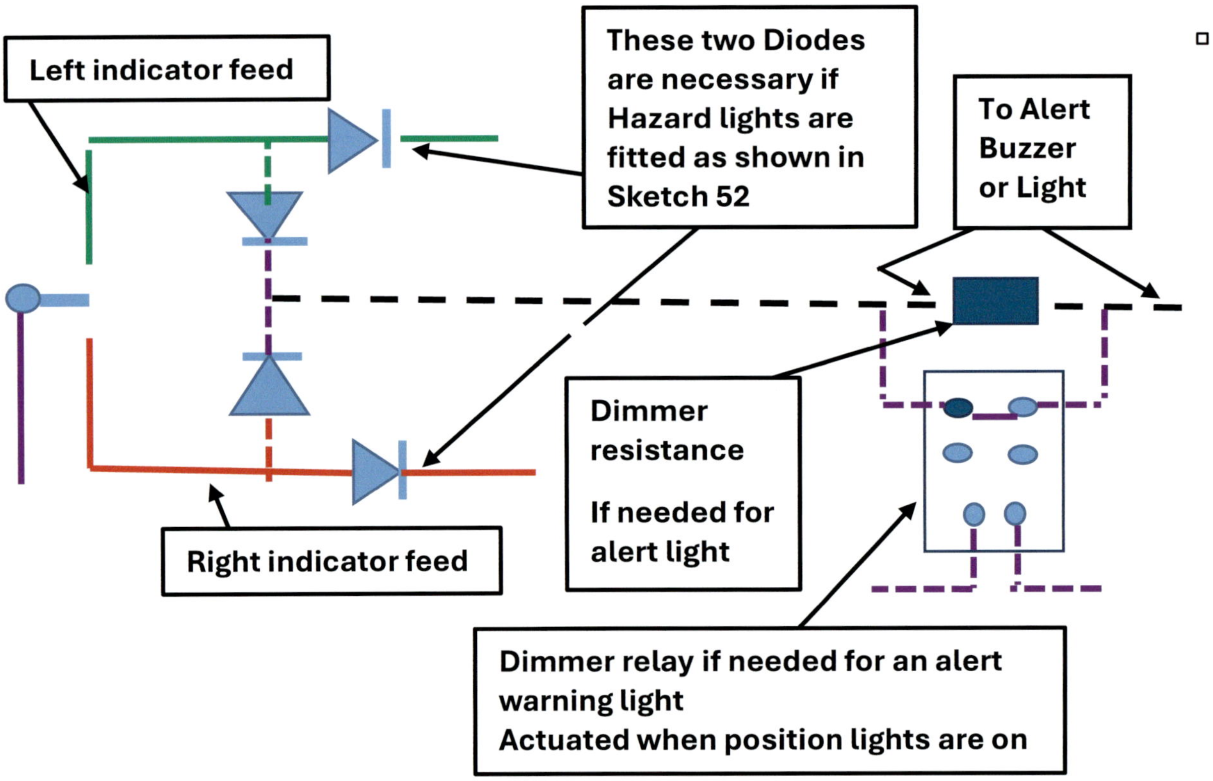

Sketch 58 – Suggested circuit for an indicator alert connection.

Photo 49 shows the extra Hazard light switch on the Mk 1 Machine on the left-hand side. The small extra warning light on the right is very useful; a battery voltage indicator connected directly to the "ignition on" battery positive which changes colour from red to yellow to green depending on battery/charging voltage. (Available from Paul Goff and well worth fitting to any bike, with two or three wheels, as an "always there instant check" to avoid a flat battery if the charging system fails.)

The Mk 2 has a second extra switch on the right-hand side for the position lights as the existing base bike circuit made using the re-wired existing switch impractical.

Photo 49 – The Mk 1 instrument panel with hazard switch added.

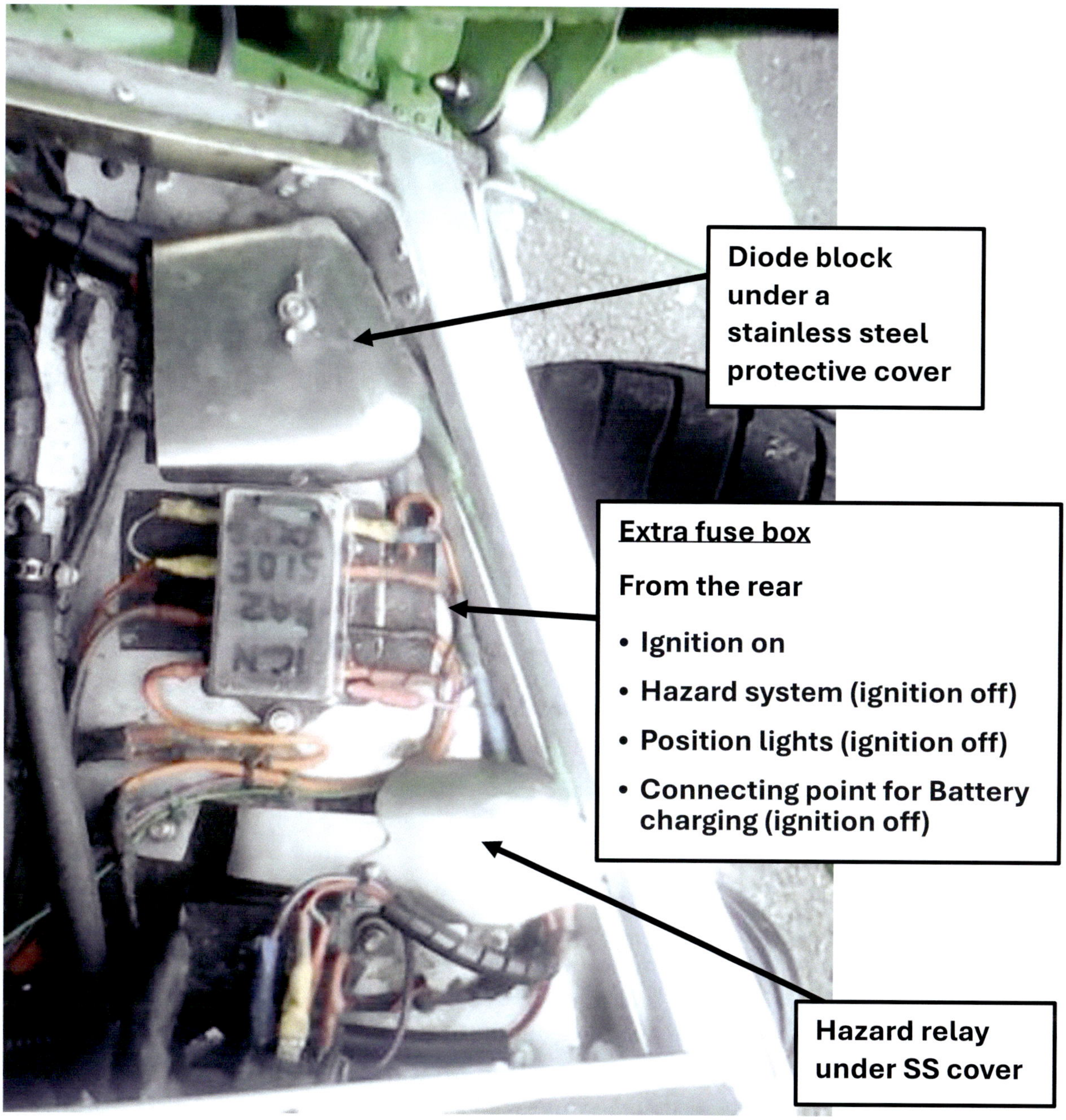

Photo 50 – The front compartment extra electrical panel on the Mk 2.5 machine. (The panel is on top of the inner wing panel in a relatively protected space which could not be conveniently used for anything else. The battery terminals are somewhat hidden, so the remote charging point is useful occasionally.)

9 – Riding Position

The objective of this type of trike is to make the riding experience as close to the two-wheeler as possible. The riding position could be left unchanged from the base bike but there are advantages in both comfort and control by making some small changes. On a two-wheeler the main requirements are to be able to resist acceleration and braking forces, move body weight as necessary and provide footrest clearance for cornering lean, while remaining firmly in position on the bike. On a three-wheeler, the acceleration and braking loads do not change much but body weight movement is much less important and lean clearance for the footrests is not needed. There are, however, other requirements, the need to resist cornering forces when upright and, with two wheels at the front, the need to be able to apply more steering force to the handlebars particularly at very low speed.

9.1 – FOOTRESTS AND HANDLEBARS

Footrests

One of the advantages of three wheels is that there is no need to provide cornering lean clearance under the footrests so they can be lower and somewhat further forward which often provides a more comfortable riding position. Also, the different loads on the rider during cornering and extending the footrests outwards provides a wider base to resist cornering loads. Moving the brake and gear pedals to suit new footrest position is usually quite easy to arrange.

All bikes are different, but Photo 51 shows the arrangement on my second machine with the footrests moved down 3 inches, forward 1 inch and outwards 2 $^{1}/_{2}$ inches relative to their original position.

This was achieved in two stages, The ZXR 400 footrests and brake master cylinder and pedal are mounted on a bolted-on carrier which was moved down by 1 inch and outwards by I/2 inch by fitting a spacer bolted to the frame bolt positions and then bolted to the footrest carrier lower down. (This stage had already been done as a two-wheeler as the standard riding position seems to have been designed for very small riders with short legs and long arms.)

The second stage was to fit new L shaped footrests which were cut from 25mm aluminium alloy plate and mounted on the original footrest lugs. These new footrests move the footrest position a further 2 inches down, 1 inch forward and 2 inches outwards.

The gear pedal was already pivoted on the footrest mounting plate on the base bike, so it only needed a longer connecting rod and the pedal extended outwards to accommodate the change.

The brake pedal was adjusted down on its original pivot and the foot pad extended outwards. The resulting small change in leverage seems acceptable in use.

Photo 51 – Lower and wider footrests on the Mk 2 machine.

Handlebars

Depending on the base bike arrangement, not much change to the handlebars is likely to be necessary, but slightly longer bars with the grips moved towards the rider to give a more upright riding position and more leverage when cornering is an advantage on the road.

The base ZXR400 bike of my second machine originally had low clip-on type handlebars which were far too low for normal road use on a two-wheeler and totally unsuitable for a 3-wheeler. I had previously replaced them with raised bars, so no change was necessary with the three-wheel conversion other than pivoting them slightly rearwards. Photo 52 shows the arrangement.

The raised bars were made from 22mm diameter 3mm thick type 316 stainless steel cut and welded to provide a step upwards so they could be fitted to the original clamps on the fork tubes. To provide clearance, it was necessary to cut away small parts of the front fairing but otherwise the new bars have proved satisfactory on both two and 3 wheels.

One advantage of three wheels is that the mirrors can be moved further outwards, no longer providing a good view of your elbows, but actually allowing you to see what is behind. Well worth doing, whatever handlebar arrangements you chose.

Photo 52 – The cranked, raised handlebars on the Mk 2 machine.

9.2 – FAILURE ANALYSIS

This brings us back to failure analysis (which should never go away) as the Mk 2 handlebars and, to some extent, the footrests are **Critical Parts**.

The handlebar material used on the Mk 2 is similar in strength to the original and is sufficiently ductile to ensure that it would always bend in an overload situation before it would break. Maximum handlebar loads are limited by rider weight and strength, but normal running loads are likely to be higher than on a two-wheeler because of heavier low speed steering and occasional single wheel bump shocks.

The cranked sections are also subject to torsional loads but have large strength reserves in this respect.

The possibility of fatigue failure at the welded positions is therefore the limiting feature. The welding was done by an expert welder with long experience in the aero-engine industry and inspected by dye penetrant crack test and an internal visual check by small diameter introscope to confirm full circumference weld penetration (old contacts can come in useful at times). These precautions, together with regular inspections were judged sufficient and no problems are anticipated.

The new footrests were covered by ensuring large strength reserves on all 1-off (2-off) parts relative to the original parts, using the same logic that loads, however they are generated, are limited by rider weight and leg strength. This together with occasional inspection was judged sufficient and there have been no problems to date.

10 – Bodywork

It is necessary to provide some form of front bodywork to comply with the construction and use requirements that there should be no sharp edges which could injure a pedestrian in an accident. I believe the individual vehicle approval requirements specify a minimum edge radius of 3mm and specify checking to show that any sharper edge cannot contact a 30cm diameter ball, but it is worth checking on the DVLA website as requirements are often updated.

Rubber beading or similar edging strips fixed with adhesive to cover sharp edges are not accepted and all bolts and other fasteners protruding from the bodywork should have domed heads.

The second thing to think about is cooling air flows to and from the radiator, oil cooler and the area in front of the engine. With the radiator moved to the front, the best way is to arrange for air passing through the radiator to mainly exit on the top surface of the bodywork above the radiator space as air flow along the length of the inside of the front compartment is usually restricted and could easily be blocked by luggage, etc. carried in this very useful space. A separate exit also helps to keep rain out of the front compartment.

Cooling to the front of the engine and the oil cooler, if it is left in its original position, is best achieved by means of an air scoop or deflector picking up cooling air from below the front floor.

A forward facing intake slot approximately 15 inches wide and ½ inch depth below the floor line on the centre line, as shown in Sketch 54 in the cooling system section, with the cooling air deflected upwards by a suitable curved deflector has proved sufficient on my machines and does not pick up too much water or road debris. If the radiator is left in its original position, additional ducting to guide the under floor air through the radiator would probably be needed.

That said, the choice of front bodywork is entirely up to the individual and what they can either make or get made.

For the registration and early road testing, I would suggest the minimum of bodywork to satisfy the sharp edge requirements, provide some sort of mudguard cover for the wheels and support the lights, etc. More elaborate bodywork is very time consuming to make, or costly to get professionally made, and could be fitted later when the steering, suspension and other road-going characteristics have been proved.

The method I have used is to produce an underframe for the front bodywork with all the attachments, hinge positions and locking features in place. Lightweight aluminium tube is a good material for the underframe, or even some sections of marine plywood, as it does not need much strength and much of the necessary stiffness will be provided by the body panelling. Sections hinged for access are best with the hinges at the front; rear hinges would necessitate car type two stage locking catches and other precautions to avoid possible lifting while on the move due to wind pressure.

Sketch 59 shows the general idea and Photo 53.1 shows the aluminium front bodywork underframe of the first machine, which has all-enveloping type front bodywork. The front bulkhead is marine plywood and the whole thing is hinged at the bottom of the front bulkhead to allow it to be opened by lifting at the back. The central strip holds the opening handle which pushes down flush when not in use, (an aircraft surplus part from long ago), and operates two hooks at either side, engaging under the fixed front sub-frame edges, (see Photo 53.2). Rubber strips along the joint compress to prevent rattles and keep water out.

The centre section, which is almost single curvature, is panelled in aluminium, pop riveted to the underframe using countersunk rivets and bonded with silicon adhesive.

The aluminium strips over the wheels provided the basic shape and were covered with fibreglass laid up on a fine metal mesh.

The front section was shaped in medium density polystyrene, bonded to the bulkhead with silicone adhesive and covered with fibreglass.

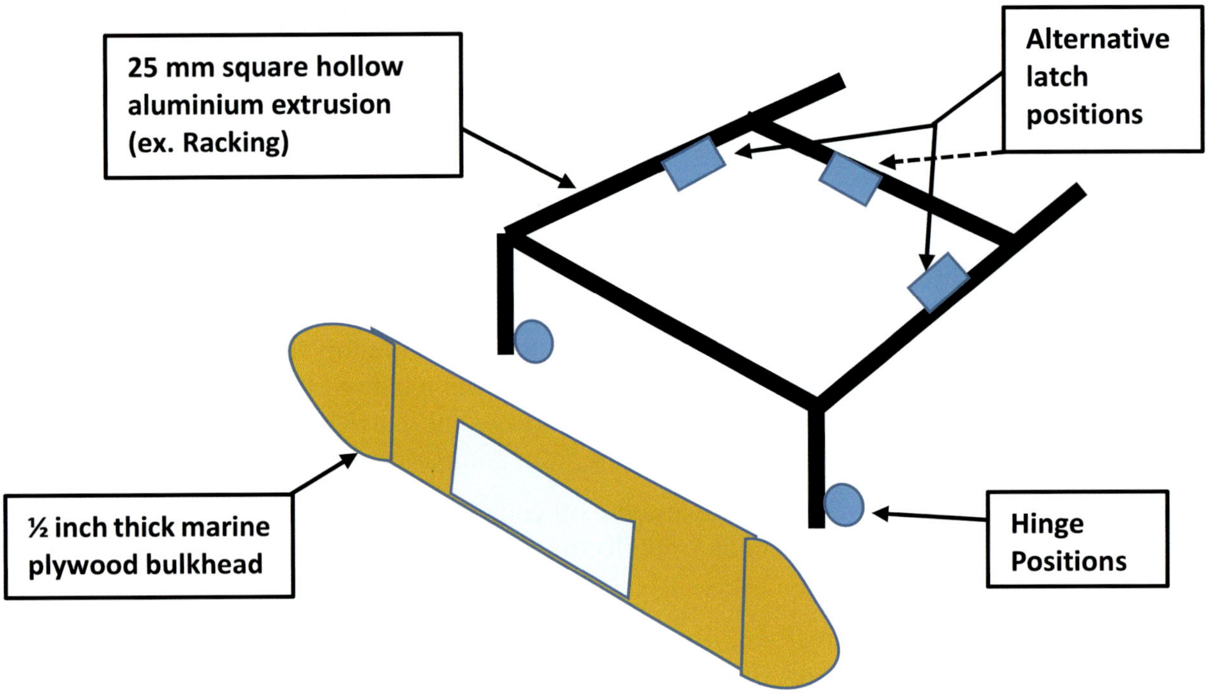

Sketch 59 – The underframe supporting the front bodywork and positioning the hinges and latches on the first machine.

The bodywork of the second machine has similar construction but is an open wheel type, with mudguards turning with the wheels, so only the nose section is fibreglass. The nose section and bulkhead are fixed to the front sub-frame and only the top section is hinged to open.

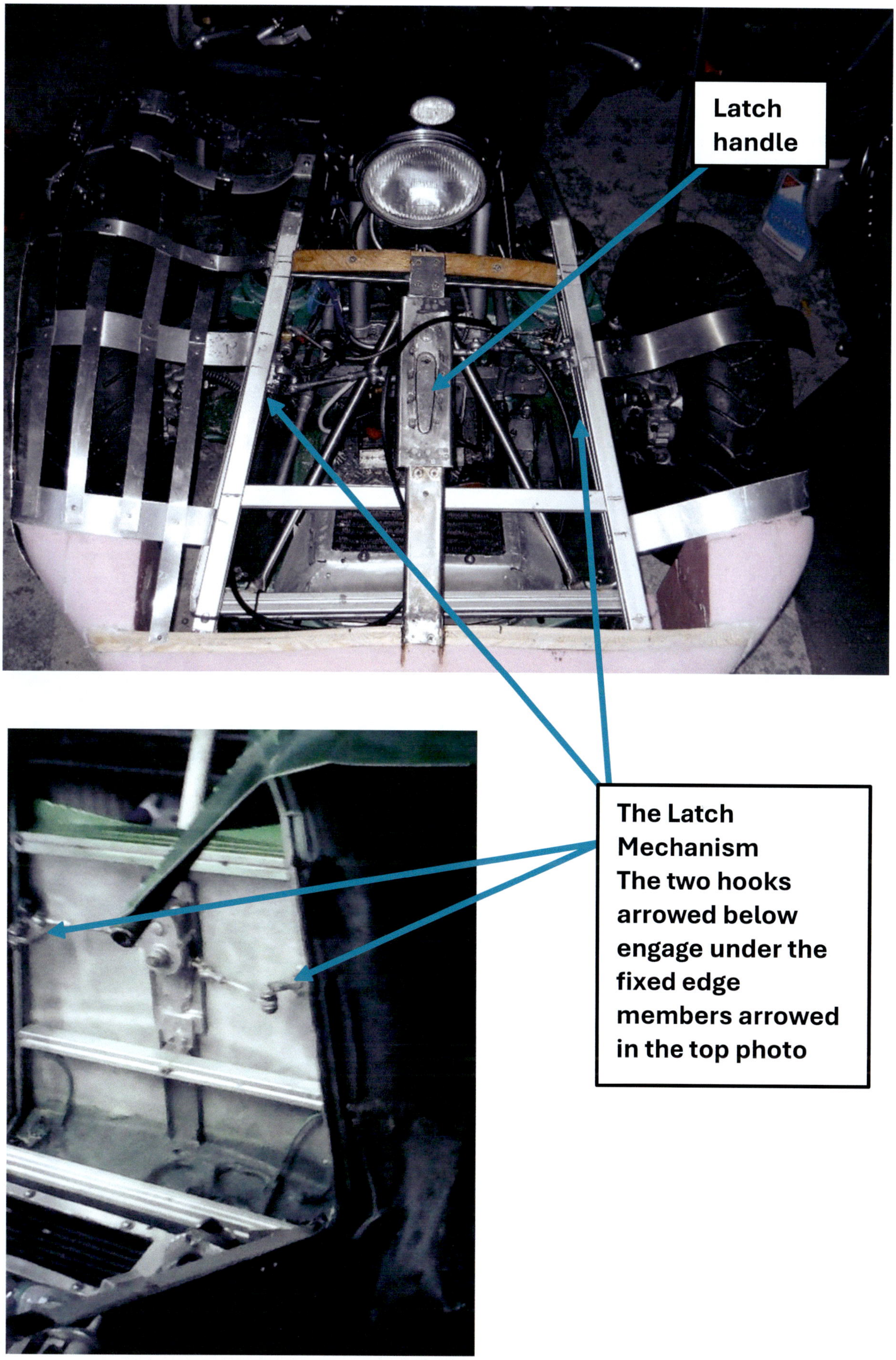

Photos 53.1 and 53.2 – The structure of the front bodywork during construction with the shaped polystyrene nose core in position. The latch mechanism on the finished bodywork is shown in the bottom photo.

The whole thing needed a lot of work, a certain amount of body filler and a professional paint job to get to the smooth finished shape but has come out well. Photo 54 shows the finished front end of the Mk 1 machine. (I am often asked where I bought the bodywork from.)

There are undoubtably lighter methods of construction which would be used on a production machine, but they require making of an accurate mould, or other costly and time consuming work and are perhaps unnecessary for a one-off.

Photo 54 – The finished front bodywork of the Mk 1 (the first machine).

11 – Bodywork Styles

The choice is between all-enveloping and minimalist or somewhere in-between and is another decision to be taken early in the project to avoid a lot of extra work. I don't believe that, at normal road-going speeds, there is much to be gained from the differences in aerodynamic drag or downforce between the different types, although low drag and some positive downforce are always advantageous.

At racing speeds both would become extremely important, particularly avoiding serious upwards front end lift on a very light-weight machine if the front end came off the ground. There has been at least one case of a racing sidecar somersaulting on a bumpy road course due to the aerodynamic lift from the flat under floor when the front end was lifted by a road bump, etc.

The weight of the bodywork is not likely to be a large proportion of the overall weight, so although saving weight is always a good thing, the bodywork weight will only move the overall centre of gravity slightly further forward. This in theory would improve straight line stability but tend to promote more cornering understeer, but the effects are likely to be small.

One thing which is important is protecting the rider as far as possible from wind pressure and rain or spray from the front wheels on wet roads.

Taking account of the above, appearance, the builder's preferences, ease of manufacture and satisfying the applicable construction and use regulations become the most important factors in deciding what sort of bodywork to make.

In my opinion, there is a lot to be said for producing some temporary minimum bodywork, just enough to satisfy the requirements for initial testing, MOT and registration, as it would save a lot of work in the long run and allow the final bodywork to be fitted to a fully roadworthy machine.

I have produced three different body-work styles for my Mk1, Mk 2, and Mk 2.5 machines, each one getting smaller, but not yet a true minimalist design (but who knows, here may be a Mk 3 still to come). Photo 54 shows the Mk1 all-enveloping design which worked well but is heavier than I would have liked and tends to look too bulky from some angles.

Photos 55 and 56 show the Mk 2 and Mk 2.5 open wheel designs in attempts to improve on the Mk 1 (Photo 54). I think it is moving in the right direction, but everyone has their own opinion.

Open wheels bring their own complications in that they need mudguards carrying the lights. I used two of the original bike front mudguards with a substantial single side mounting made from 50mm x 10mm aluminium alloy strip heated and bent to shape. The support is attached to the swivel hub carrier at existing steering arm and brake calliper bolt positions. This arrangement has proved satisfactory with no additional support. The rear lights, indicators and brake lights are mounted on the mudguards with the front position lights mounted on the hinged top section. Photo 57 shows the arrangement on both the Mk 2 and Mk 2.5 machines.

The Mk 2.5 is the same machine as the Mk 2 except for the front end of the bodywork. I liked the appearance of the open wheels and cycle wings but the Mk 2 full width nose section, to me, looked too bulky in relation to the rest of the machine.

The end sections were simply cut off vertically and replaced with two small wings. These are shaped to provide some downforce as a matter of principle although, in practice at road speeds, the effect is small.

(In theory, the downforce created would approximately balance the front end lift at higher speeds as a result of aerodynamic drag and, as both drag and downforce will increase with the square of the speed, all other things being equal, a fixed incidence angle would be somewhere near optimum over the whole speed range.)

I think the Mk 2.5 looks more balanced than the Mk 2 in appearance and it saved approximately 20 lb in weight, but it did necessitate changes the ends of the hinged top section.

Photo 55 – The Mk 2 front bodywork.

One change always leads to others so getting it right first time is always best.

The triangular white flaps in front of the position lights on the Mk 2 move down before the hinged top panel is lifted. A necessary but troublesome complication as the only possible front hinge positions caused the front edge of the hinged top section to move forward at the outer ends during lifting and would otherwise foul the edge of a one-piece nose section. The Mk 2.5 changes remove the need for these flaps.

Photo 56 – The Mk 2.5 front bodywork.

With the removal of the ends from the Mk 2 nose section, the ends of the hinged top had to be trimmed and the position lights moved to the edges. The additional position strip lights on the front edge of the top panel conceal the exposed flat edge (a bit of a bodge job but at least it got rid of the troublesome drop-down front triangles see Photo 55). Perhaps this area will be improved in time. Photo 25 shows an angled view of the Mk 2.5 showing the extent of the bodywork reduction.

Photo 57 – Showing the front mudguard arrangements on the open wheel machines.

Finally, there is no substitute for a good paint job to enhance the appearance of the exposed parts of the new front body work. Both the design of the paint scheme and a professional finish make a lot of difference to final appearance of the machine. An important factor in any unique project.

(If someone asks you where you bought the body work, as they often do, you will know you have done a good job.)

12 – UK Registration

There are basically two routes to UK re-registration of a two wheeled bike converted to a three-wheeler. You could say, the “easy way” and the “hard way”, and it depends on the degree of change to the original vehicle.

Basically, if the bike already has a registration document and number plate, you need to keep as much as possible of the original bike, unaltered, and just add a new front end. Then hopefully the “easy way” is open to you.

The detailed requirements seem to keep changing so you should refer to the DVLA web site www.gov.uk/vehicle-registration/kitconverted-vehicles and go to the section called “Radically altered vehicles”.

The following is based on my experience in re-registering two conversions and is worth bearing in mind from the start of the project. Then a few months before the likely completion date, I suggest that you should write to the DVLA at:

Kits and Rebuilds

D9

DVLA

Swansea

SA99 IZZ

Just say that the bike, registration number --------, is being converted into a trike with minimum change to the existing machine and ask them what they require in order to re-register.

Unlike some parts of the DVLA, this department seems to be staffed by interested and helpful people and they will reply providing a list of their requirements.

That said, my experience of the system is as follows:

The rules for keeping the same number plate are a good guide to an acceptable level of change in order to use the easy way to re-registration and they are based on a points system with a minimum of 8 points retained unchanged.

- *The frame with no welding or other radical change counts for 5 points*
- *The suspension, both ends, counts for 2 points*
- *The axles, both ends count for 2 points*
- *The transmission counts for 2 points*
- *The steering counts for 2 points*
- *The engine counts for 1 point*

With two wheels at the front, the front suspension, front axles, and steering must change so that leaves the frame, transmission and engine, a total of 8 points, just enough if you avoid significant change to the other three. That is don’t change the engine or transmission and just bolt an additional front structure to the existing frame, avoiding welded attachments.

Having got to this point, the process is likely to require submission to the DVLA of the following.

- *The two-wheeled registration document.*
- *An MOT certificate in the three-wheeled form.*
- *A weight certificate from a certified weigh bridge.*
- *Completed DVLA form V627/1, which provides a description of all aspects of the change. showing why it is acceptable.*
- *Evidence of type approval (if required).*

The first 3 are relatively easy to provide except that very few MOT stations are approved for Class 3 vehicles (3-wheelers) and you need to get the machine to and from the station on a trailer or get it collected and returned. This can get expensive and time consuming, especially if the MOT shows some points which do not comply with obscure three-wheeled MOT rules.

The weight certificate is needed to show which taxation class it falls into although it will definitely be a light three-wheeled vehicle. It can be done on the MOT brake test machine, if you ask nicely, but you will probably need to provide your own blank certificate and get them to fill it in and put their MOT stamp on it. Make sure they measure the weights separately, as MOT brake tests weights normally include the tester's weight. Attachment 2 shows the certificate I used which was accepted without question.

Form V627/1 is where you are required to supply details of the conversion. It provides a series of aspects of the conversion each of which need to be answered. Receipts for any major new parts are required although most of my parts were from my own stock accumulated over many years. There is very little room on the form for details, so it is better to just attach an additional sheet with room for the required information and just refer to it on the DVLA form. A few photographs of the complete machine showing the unchanged base bike are also useful.

Attachment 1 shows the additional note I added to the Mk 2 application, which was accepted without question, so seems to have been what is required. It should give an idea of the sort of information they need but every conversion will be different.

The sting in the tail is Evidence of Type Approval if required, but in the case of minimum change, a photograph of the VIN plate in position and unchanged seems to be sufficient. Older machines prior to the introduction of type approval presumably would not need to cover this requirement.

Significant changes to the existing bike at the same time as re-registration as a three-wheeler, are likely to lead to the "hard way", which will require an IVA (Individual Vehicle Approval) test at a government test centre.

There are a separate set of IVA construction and use rules, available online and worth taking into account, even if you are aiming for the "easy way" to re-registration.

This test is designed to ensure that the machine complies with all these rules and does not replace the normal MOT test. I have no personal experience of the test, but the IVA rules are very clear and, if taken into account from the start of the project, it should not be too difficult to gain approval. It is, however, much more expensive with test fees and the need to trailer the machine to some remote test station, perhaps more than once as many things are subject to the tester's judgement and there is no scope for argument.

13 – Final Thoughts

I think that the final result is worth all the time and effort put into the job but that is for others to judge.

This type of conversion can only be done with plenty of time available, somewhere warm and dry to work in and a good selection of tools and equipment available, (a few good engineering contacts also helps).

Without the above, it would be better to go for a normal trike conversion with two at the rear, which would require much less work and development testing and could be largely built from ready-made components bolted together.

I hope that the book encourages a few people to have a go at a conversion, which ever end they choose, as a unique trike, once up and running, can provide a special riding experience unlike any standard manufacturer's product while retaining all the maintenance and reliability advantages of the well proved manufacturer's product.

The two at the front arrangement, once you get used to its peculiarities, has certainly proved to result in a very entertaining and confidence building ride with very controllable cornering where loss of tyre adhesion at either end in slippery conditions can be easily recovered or even used to assist cornering.

In straight line running, the feel is very similar to a two-wheeler, except that the machine leans with road camber changes while continuing to run straight, which feels a bit odd at first to a two wheel rider were lean means turn and you need to learn to ignore lean changes during straight line running.

Braking is also good and confidence building, without ABS, as momentary wheel locking causes no problem in control. They are, however, sensitive to balance between the two front brakes as more braking on one side can only be resisted by the rider through the handlebars. As long as both callipers are in good condition it is not a problem, but it soon tells you if one side is becoming less effective.

Very rough road surfaces are the most difficult, in general requiring a steering damper to counter momentary deflection of the steering, and the best compromise you can find for the front suspension.

Finally, best wishes to anyone who attempts this type of conversion or any modifications to a motorcycle. I hope your efforts are successful and provide many hours of entertaining building and riding.

(Just don't forget the **Failure Analysis** to keep your efforts safe.)

Attachment 1

DVLA Form V627/1 Example

Motorcycle Registration Number ------- converted to a Tricycle

This change to a motorcycle, which I have owned since new in year 2000, has been designed and constructed by myself, (a mechanical engineer with over 60 years' experience in aircraft and modified road vehicle engineering).
The intention of the modification is to convert a motorcycle into a "two at the front" tricycle using car-based steering and suspension components. The motorcycle front wheel and brake are retained and a second matching wheel added.

General The whole of the existing motorcycle is retained with the exception of the lower sections of the front forks and rear brake calliper, with minimal changes necessary to comply with construction and use requirements.

Structure No modifications have been made to the existing motorcycle frame. The front sub-frame is attached by means of 11 tubular struts bolted to existing strong points on the frame in a triangulated structure similar in principle to the method of attaching a bolted-on side car.

Steering BL Metro steering arms at the wheels are operated by means of two track rods connected to the existing motorcycle handlebar and front fork assembly using Metro/Mini type ball joints.

Brakes The existing motorcycle front and rear brakes are both retained unchanged. A second identical motorcycle front wheel with the same type of brake is added. A mechanical handbrake has been added by using a new, car type, rear calliper, with the same piston area, from a Citroen Saxo, operating on the existing rear wheel disc. This calliper is the second of a pair obtained for the similar conversion of the Honda motorcycle, registration number -------.

Suspension The existing motorcycle rear suspension is unchanged. The front suspension uses Rover Metro and Mini components adapted to suit the motorcycle front wheels and discs. The suspension units are motorcycle units similar to the existing rear suspension unit.

Front fairing and lights A full width front fairing and top fairing are incorporated, aligned with the existing motorcycle fairing panels, and providing a smooth exterior surface with no sharp edges The existing motorcycle front mud guard is retained with an identical second mud guard covering the front wheels. All additional obligatory lights are mounted on the top panel and mud guards with the existing motorcycle real lights also retained.

Colour The existing green and white paint on the motorcycle is retained. The front panels and mudguards are the same colours.

Minor parts removed for the modification will be retained in case of possible future re-conversion to a two-wheeled motorcycle.

The whole of the existing motorcycle, with the exception of a few minor parts, is retained unchanged including the engine, frame, transmission and rear axle assembly, justifying retention of the existing registration number.

The front sub-frame, complete with suspension, and steering is constructed using car and motorcycle components, largely from my own stocks, spare from previous projects and all have been fully reconditioned, (receipts attached).

Attachment 2

Suggested weight certificate to be filled in and stamped by MOT tester

If a blank certificate is supplied, with luck the MOT station will fill it in

MOT Station Headed Paper

Date ------------

Tricycle Reg. No ----------- VIN No. --------------

Front Axle ------------- Kg

Rear Axle ------------- Kg

Unladen weight ------------ Kg

Vehicle weighed at time of MOT on calibrated MOT brake test equipment

MOT Station No. ------------------

Signed ---------------- MOT Tester

MOT Station Stamp

Acknowledgements

With thanks to all who have helped with my various efforts with time, materials, parts or ideas over many years including Derek and Simon Green, and to Andrew Thompson for his superior paint work on all my recent efforts, Mark Woodward for publishing advice and for his efforts to keep us up to date with the evolving motorcycling scene through the Norton Owners Club magazine.

Also, thanks to Paul Adams for use of his Vincent photos and to Paul Kneen for his help with all things related to my many visits to the Isle of Man.

Also, to my wife Jen for tolerating (not without complaint) the occasional smells of curing fibreglass, paint, solvents, petrol, diesel, exhaust gas, etc. and the occasional loud noises which are an inevitable part of special building.

Appendix 1
Example of Aero Engine Type Analysis

A typical piece of aircraft engine type failure analysis to show the sort of work involved. The analysis is applied, for interest, to a typical motorcycle telescopic front fork.

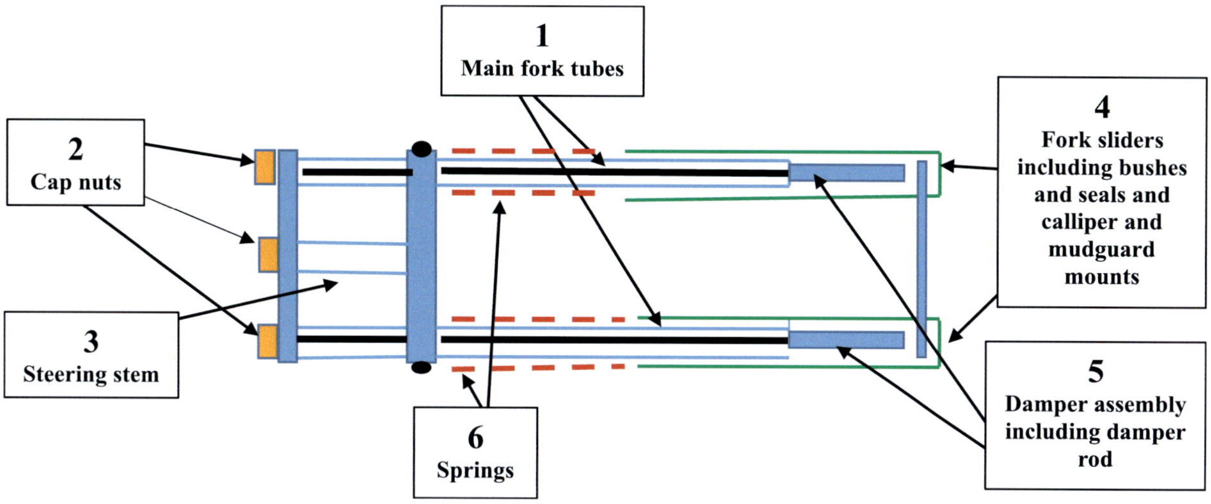

Sketch 60 – Typical telescopic forks.

Failure Categories

- **Minor** – Would not affect continued running.
- **Major** – Would affect continued running and would require increased rider skill to ensure a safe outcome.
- **Hazardous** – Outcome could not be controlled by rider. Could result in death or injury of the rider or another person.

Note these categories have been adapted from the strict aircraft definitions to suit the application to a motorcycle.

Component No 1 – Main fork tubular leg (Two)				
Failure mode	**Effect**	**Justification for safe operation**	**Probability** **(Note: a numerical probability would have to be justified if the failure category is Major or Hazardous.)**	**Failure Category**
Minor fatigue cracking or material defect	No effect on operation but would not be detected and could progress to more severe cracking	No effect on safe operation provided that it did not progress to more severe cracking (below)	Minor surface cracks or defects are unlikely to progress to the point of failure in the life of the machine if the fork tubes are from a production machine or are of equivalent material and manufacturing quality, and experience similar stress levels in operation, justifying a satisfactory probability of failure. {Documented experience in service required}	**Minor**

			[Stress statement required]	
Severe fatigue cracking	May not be detected particularly if the fork tubes are shrouded. Failure of a fork tube could occur, probably in a high load situation such as riding over a severe pothole at speed. **The fork tubes are therefore Critical parts**	<u>If the fork leg is a standard part proven by use on other bikes:</u> Supporting evidence covering the following is required. Origin of fork legs Undamaged condition Comparison of intended use with that of standard part. <u>If the fork leg is a unique part</u>, The following aspects need to be shown to be satisfactory and fully documented. Strength and fatigue life {Stress statement required} Material quality {Control of source and specification of material to ensure consistent material quality and strength.} Manufacturing methods {Control of the manufacturing process and any specified testing to ensure that the product conforms to the design intent.}	In the case of a standard part, a numerical probability could be based on the estimated total number of miles covered without problems on all the production machines of this or a similar type of the forks and similar or greater weight. [Documented experience in service required. Stress statement required.] The probability of failure of a unique part could be estimated from the material manufacturer's published fatigue life properties for the material, or by comparison of stress levels with other similar satisfactory parts in general use. {Probability statement required}	**Hazardous**
Minor corrosion	If corrosion occurred, it is likely to extend to the sealing surfaces of the damper fluid seals resulting in fluid leakage.	Would most likely be detected by damper fluid leaks from the seals around the fork tubes before any fatigue cracking could develop.	Corrosion is likely to occur in the worst environmental conditions but would be detected before severe fatigue cracking could develop. Internal corrosion will not occur as tubes are sealed against ingress of moisture.	**Minor**

Component 2 – Fork tube and steering stem cap nuts				
Loose or lost top cap nuts at top yoke position	One loose nut would have no noticeable effect. Lost cap nuts are unlikely as the damper rod is attached to the nut. If both nuts were loose, there may be some extra flexibility in the forks but unlikely to be sufficient to be detected.	The main running loads are supported by the bottom yoke so strength would not be significantly affected.	A loose cap nut could occur, although unlikely, but the effects would be Minor.	**Minor**
Loose or lost steering stem cap nut	Loss of end load on the steering head bearings resulting in reduced location of the forks which would be detected by movement in the forks during braking.	Loose steering head bearings would be detected before they could have a significant effect on safety.	A loose cap nut could occur, although unlikely. The effects would be most likely to be **Minor,** but could be **Major** if the increased fork flexibility was sufficient to require some additional riding skill to maintain control. If judged to be **Major**, the numerical probability could be based on the record of the very large number of production machines with the similar design, screw thread type, tightening torque and thread locking arrangements. {Probability justification document required}	**Minor** or possibly **Major**
Component 3 – Steering Stem and Steering Head Bearings				

No need to go any further. The above shows the type of work required to produce a formal Safety Analysis for an aero engine. It is easy to see that, by the time evidence to support the justifications and probabilities has been established and documented, quite a lot of effort would be needed to cover just these two components and there are also the sliders, the yokes, spring, steering stem and bearings, dampers, wheel spindle, and the mudguard and calliper attachments still to be covered before the whole fork assembly is shown to be satisfactory.

All this work is not practical or necessary for a one-off motorcycle, but the principles of the method are simple and straight forward and definitely worth a bit of thought when doing any changes to a production machine or doing something unique.